Kesorn Pechrach

Arc Control in Circuit Breakers

Low Contact Velocity

2nd Edition

Kesorn Pechrach

Arc Control in Circuit Breakers

Low Contact Velocity

2nd Edition

Pechrach Publishing

Arc Control in Circuit Breakers
Low Contact Velocity
2nd Edition
By
Dr Kesorn Pechrach

ISBN 978-0-9931178-7-9

PECHRACH PUBLISHING
7 Boundary Road, Bishops Stortford, Hertfordshire, CM23 5LE, England, United
Kingdom. Tel: (+44) 1279 508020, +44(0) 7779913907 and +44(0)7443426937

Published 2017 by Pechrach Publishing

Introduction

This 2nd Edition of the Arc Control in Circuit Breakers: Low Contact Velocity, the new innovation data, discussion and analysis the insight of the spectrum, electron energy, Mass and temperature rise for each chemical element. The materials and condition of the Arc chamber: Polycarbonate and ceramic, Contact materials: Ag/C step and Cu punch, and Contact opening velocity: 1 m/s and 10 m/s, have been observed.

Dr Kesorn Pechrach
1 January 2017
England

Acknowledgments

An ocean of thanks to:

My awesome family in Thailand for their continuous support, understanding and encouragement. Especially, my beloved sister Somjan Pechrach for donating one of her kidneys to me while I was writing this book.

My talented supervisor, Prof. J.W.McBride for firing, shaping and creating a rare diamond Dr. KESORN PECHRACH out of the mixing earthen and stoneware.

My brilliant consultant and amazing husband, Dr. P.M.Weaver for enthusiasm, advice, suggestion and non-stop support.

Special Thanks to my son, Neran J. P. Weaver for questions that I cannot answer.

My best friends P.G. Wheeler and P.W. Wilkes for their cheerful supply and manufacturing of mechanical and electrical parts for experiments (I did a lot of damage).

Prof. N. Ben Jemaa from University of Rennes 1, France, Prof. Koichiro Sawa from Keio University, Japan, Dr. J. Swingler and Dr. Suleiman M Sharkh from University of Southampton, UK for their suggestions and recommendations throughout this work.

This book is dedicated to W. Hunt, who always never ever stop.

TABLE OF CONTENTS

TABLE OF FIGURES

LIST OF TABLES

Nomenclature

B   magnetic flux density

μ_0 Permeability of free space ($4\pi \times 10^{-7}$ H/m)

γ specific heat ratio $C_P/C_V = 1.4$ (air)

r   magnitude of distance from current moment to point S at which magnetic effect was evaluated

i   arc current

j   current density

a   contact gap

V_A Arc voltage

σ electrical conductivity

E   electric field

P_0 pressure at the shock front

P_1 pressure in front of the arc

T_0 temperature in front of the shock wave

T_1 temperature behind the shock wave

a_0 sound speed

l   arc length

F_{mag} magnetic forces

F_g gas dynamic forces

D_{arc} arc diameter

P_s stagnation pressure

M   Mach number

K   Nozzle flow constant gas

$\wp_s$ density of contact material

h   enthalpy of metal vapour (overheating energies, melting, volatilization and ionization)

R_0 Ideal gases constant

c_v specific heat at constant volume

c_p specific heat at constant pressure

T_s starting point temperature

T_B boiling point temperature

T_m melting point temperature

T_G gas temperature

J the total current density,

J_i the ion current density,

J_{em} the electron current density

J_{ed} the back-diffusion electron current density

P_{arc} arc power

CHAPTER 1

INTRODUCTION

1.1 Background

In the low and medium voltage range, a circuit interrupting device is used to interrupt prospective peak short circuit current up to 100,000 A. These devices must have the ability not only to interrupt load currents, but also to interrupt a short circuit when the fault current can reach a magnitude many times full load.

The fuse, by comparison with circuit breakers, suffers the disadvantage that replacement is necessary after operation. The physical size of the fuse and therefore, its cost, is directly proportional to its current rating. The fuse, being a thermal device, generates more heat than the current carrying parts of a circuit breaker of equivalent normal load.

The circuit breaker is of vital importance as a device used for making and breaking an electrical circuit under conditions of varying severity. The functions are [1]:

- It must be capable of closing and carrying full load currents for long periods.
- Under prescribed conditions, it must open automatically to disconnect the load or some small overload.
- It must successfully and rapidly interrupt the heavy currents, which flow when a short circuit occurs.
- With its contacts open, the gap must withstand the circuit voltage.
- It must be capable of closing on to a circuit in which a fault exists and immediately re-opening to clear a fault from the system.

- It must be capable of carrying current of short circuit magnitude until, and for such time as, the fault is cleared by another breaker nearer to the point of fault.
- It must be capable of withstanding the effects of arcing at its contacts and electro-magnetic forces and thermal conditions which arise due to the passage if currents of short circuit magnitude.

In circuit breakers, the arc exists in a mixture of air, nitrogen, oxygen and metallic vapor. Interruption is due to elongation of the arc, which results in cooling, and de-ionization by diffusion. Owing to the high temperature of the arc relative to the surrounding air, the arc is subjected to strong convection currents, which coupled with the electromagnetic effect of the current loop, causing the arc to move.

The subject of the electrical arc is of interest from the theoretical viewpoint and also of considerable practical importance. The design of circuit breakers is mainly based on experience rather than precise science. Empirical formulae can be used to determine dimensions of certain general types and the breaking capacity rating. There has been a noticeable lack of co-ordination between theoretical and practical work. There has been no lack of experimental work on this subject, but the bulk of this work has referred to problems of a scientific rather than a practical nature.

1.2 Short circuit current

An overcurrent is a current flow more than the rated of current of the equipment. This may result from equipment overload or the failure of a component. This could cause insulators to fail. In a short circuit current, there is a very high magnitude of overcurrent from a fault of negligible impedance between conductors having a difference in potential under normal operating conditions. The conductors and insulators could melt and vaporise immediately. Additionally, the magnetic forces from high short circuit current can damage both circuits and circuit breakers [2].

When a short circuit occurs, the current flow through the circuit rises up rapidly and continues to the peak current of the AC. A natural current zero occurs every 10 ms for a 50 Hz cycle, if there is no protection circuit. This peak current is called the **Prospective Peak Short circuit current (I_{ppscc})** as shown in Figure 1.1.

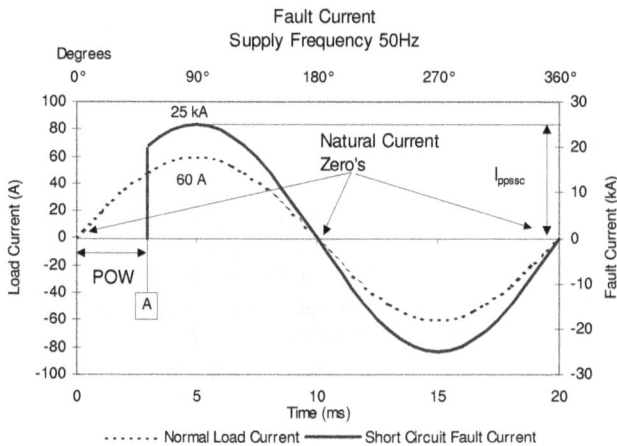

Figure 1.1: The Point on Wave of a short circuit current. Zero time is t = 0μs and is a natural current zero. The 60A peak load current is a typical value for a light industrial circuit. [A] marks the beginning of a short circuit fault [2]

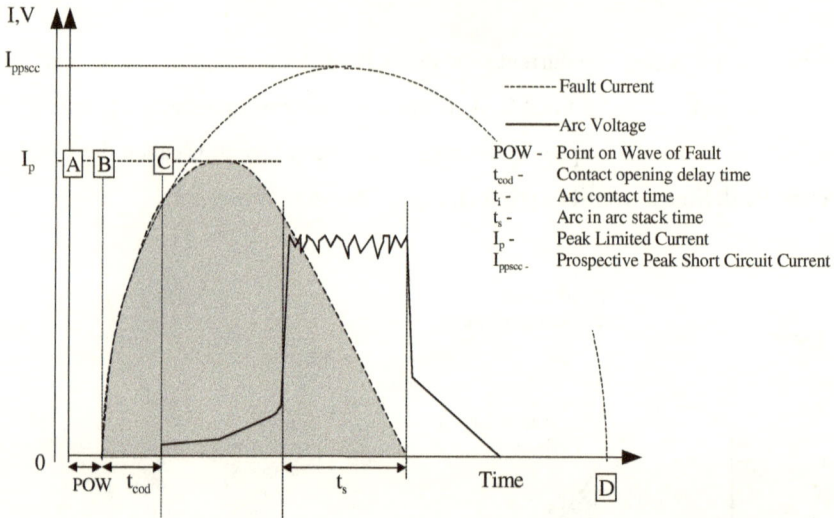

Figure 1.2: Peak Limited Current and Contact Opening Delay. The natural current zero of the supply occurs at [A], the short circuit fault occurs at [B], and the contacts open at [C]. [D] denotes the natural current zero [2]

When there is a miniature circuit breaker (MCB), the peak short circuit current is reduced. The peak current that flows through the circuit is **the Peak Limited current (I_p)**. The point that the short circuit occurs on the AC cycle is call the **Point on Wave (POW)**. The contacts do not open immediately after the short circuit current occurs, there is a delay for milliseconds before starting to open. This is called **Contact Opening Delay (t_{cod})**. Figure 1.2 shows the POW and t_{cod} relative to the natural current zero.

1.3 The operation of MCBs

The basic technique is to assist interruption by increasing plasma resistance. The arc is drawn between the opening contacts when a short circuit current fault occurs. The current through the conductors of the miniature circuit breaker generates a magnetic field in the arc chamber which acts to force the arc away from the contact region along arc runners and into an arc stack. The arc is then split by a number of parallel plates into a number of series arcs. This results in a high voltage across the circuit as shown in Figure 1.2. The arc voltage works against the system voltage to drive the current towards zero. The result of this is a decrease in the peak current passing through the circuit and reduction in the total admitted to the circuit [1,3,4,5,6,7,8].

The construction of the circuit breaker has been differently and extensively modified to suit various applications. However, the basic components are still distinguishable as shown in Figure 1.3, the cross section through a typical circuit breaker up to 400 V.

The important components of a miniature circuit breaker are the fixed contact, moving contact, arc runners, arc stack and magnetic solenoid trip mechanism. Silver graphite contacts are widely used because they prevent welding after an extended period of resistance heating. When overload occurs, the bimetallic strip heats up and trips the contact over a period of 1-120 seconds. When a short circuit fault occurs, the magnetic solenoid trip mechanism opens the contacts by accelerating a hammer. As the solenoid is energised, the contact opening delay will vary from 0.1-1.0 ms. When a short circuit current occurs, the core of the solenoid is drawn into the coil and accelerated a hammer to trip the contact mechanism. The solenoid is energised by the short circuit current itself.

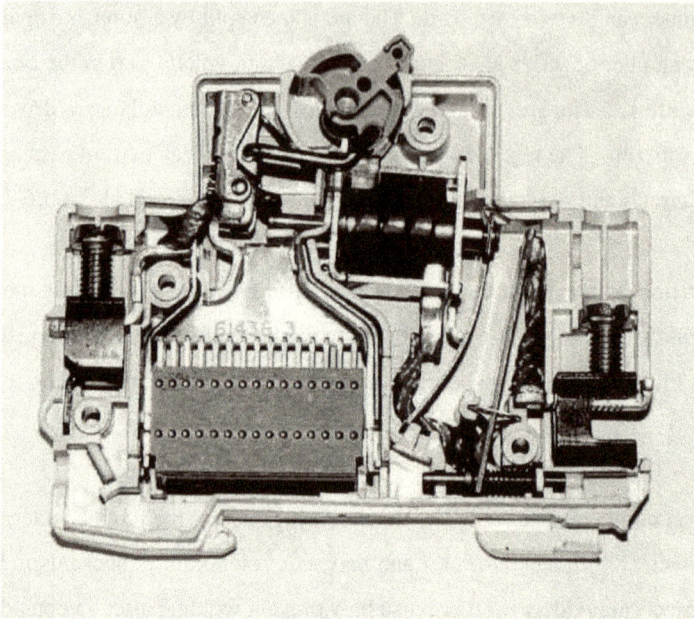

a	Tripping units	f	Arc stack
b	Switching latching system	g	Conductor connections
c	Fixed contact	h	Electromagnetic over-current trip
d	Moving contact	I	Thermal over-current trip
e	Arc chamber	j	Enclosure

Figure 1.3: The structure of miniature circuit breaker

The performance of the contact is influenced by the properties of the contact material during the switching process. The major concerns for switching devices are contact erosion and contact welding. The contact erosion occurs when the arc roots heat the contact material to boiling point. The amount of erosion depends on circuit breaker, arcing time, contact material, contact opening velocity, arc motion on the contact, gas flow and insulating materials on the arc chamber, etc.

The actual contact material is applied to a base metal e.g. copper, silver, and suitable for most requirements. Not only the electrical and thermal, but also the contact mechanical properties can be varied. Electrically, the contact materials must permit a low contact resistance, high electrical and thermal conduction. Mechanically, contact materials must have appropriate hardness, impact ductility and abrasion resistance. Contact material requirements include characteristic such as low tendency of contaminating surface layers, low burning tendency, high melting point and easy to work [6,7,8].

The arc runner is designed to direct the arc from the point of initiation to the position of final extinction along the runners. The arc stack is comprised a number of bare-metal plates with spacers between the plates to allow the arc to be split into a number of series arcs. The voltage drop of an arc on steel plates when a short circuit occurs is approximately 30 volts. For a number of N plates, the mean arc voltage in the arc stack is approximately 30N Volts.

The dimensions of the conductors system can be determined by considering the ability to carry the short circuit current. The temperature rises caused by the heating of components, contact welding and the electromechanical forces on the conductor system are also considerations.

1.4 The Research work

This is a brief review which leads to a list of objective in section 1.6 and a full review is in chapter 2. This book presents new experimental results from a test system design to recreate the current limiting operation of a Miniature Circuit Breaker (MCB).

A high speed Arc Imaging System (AIS) was used to record data about arc root commutation from the contact region. A Flexible Test Apparatus (FTA) was used to simulate the current limiting operation under controlled experimental condition. The system monitors arc motion at 1,000,000 images per second.

Computational manipulation of the optical data permits the identification of parameters describing the motion of both cathode and anode arc roots on fixed and moving contacts.

Additionally, the movement of the arc root moves away from contact region into the arc stack has been studied in detail using the arc contour image. This technique was developed by Weaver, McBride and Jeffery [9,10,11,12,13,14] and allows the arc to be viewed as a simulated arc movie.

There have been numerous investigations into the arc motion. The cathode root was thought to dominate the arc motion since the cathode is the electron source. However, there are cases where the motion of the anode root can dominate arc motion and therefore the performance of low voltage switching devices [14].

Circuit breaker performance is dependent on rapid opening of the contacts due to its effect on arc root mobility. Typical contact velocities in a commercial device when a solenoid is energised by the fault current range from 6 m/s [2]. The contact opening velocity of 6 m/s is a critical value above which parameters such as contact material, contact gap and further increases in contact opening velocity have little effect [15].

A tripping mechanism of the Flexible Test Apparatus (FTA) was developed that allowed the velocity of the contact to be varied from 1 m/s to 10 m/s. Improvements in arc control were made to permit operation at lower contact opening velocities. Low contact velocity has the potential to reduce the complexity of the trip mechanism, and make it more reliable, the mechanisms employing advanced active material actuation.

Gassing walls may effectively extend the initial time of reduced arc motion [16,17]. The composition of the arc chamber gases affects circuit breaker performance primarily by its effect on dielectric properties within the arc chamber. The ablating materials result in a faster de-ionisation of the plasma. Novel data on the gas composition in the arc chamber was obtained by optical fibre spectrometer techniques. The influence of arc chamber material, contact material and contact opening speed on arc root mobility during contact opening were investigated. In particular, interactions between contact opening velocity and arc spectrum were identified. The effects of gas composition on arc mobility are less well known.

The arc root contact time and gas flow in the arc chamber investigation were subsequently tested using fibre optic and piezo resistive differential pressure transducers. In previous work [11,18] into the arc motion it has been identified that the pressure in the arc chamber is dependent on the cross-sectional area. A wider chamber reduced the immobility time. Pressure effects on arc motion indicated that the period of arc immobility in the contact region coincided with a period of constant pressure.

There is little work about the relationship between the arc root contact time at the point at which the arc root moves off from contact region and the pressure in the arc chamber. The comparison between the pressure in the gap behind the moving contact, in the fixed contact area and behind the arc stack are presented individually for the first time for short circuit arcs ignited between low (1 m/s) and high (10 m/s) contact opening velocities. These were also analysed with optical data of the arc root moving from contact region.

The increasing performance of computers allows us to model more and more details of the switching arc. These models could reduce the time for the development of new switching devices. A semi-empirical model of the arc mobility in the contact region was developed [11,19,20].

In this book, the magnetic forces on both the arc root and arc column including the influence of the distribution of current density in the conductor were modelled. The magnetic forces in the contact area were simulated using MATLAB computer software. This model was used to calculate new data on magnetic forces as a function of arc current and contact gap. The modeling of magnetic forces can be improved the predictive capability. The purpose of validation of the model is to improve our understanding of the physical principles governing arc motion.

The movement of the arc in the contact region is governed by a combination of magnetic and gas dynamic forces [11,21]. The influence of gas dynamic force on the arc root motion is modelled individually on the cathode and anode. These models are used to interpret and explain the experimental data obtained from optical recording and pressure measurements in the arc chamber. The computer modelling results are analysed and compared with experimental outputs. An important feature of the work reported in this book is the significance of the pressure in the gap behind the moving contact.

Typical MBCs in service will be conducting current when the system is energized. The contacts are closed and a continuous current goes through the breaker. This results in Joule heating of the current conducting components, such as the contacts, the trip units, etc. The breaker therefore has to dissipate effectively the heat that is generated within it by these components [22]. At the point at which the arc root moves off from the contact region, it could be considered as a short arc. Most of the energy is dissipated in the contact area. The contact material is heated, melted, vaporized and ionized.

The resulting gas expands generating a thermally driven flow through the arc chamber. The balance between this flow and the Lorentz force on the arc is a governing factor in the arc immobility [11,23].

A new approach is proposed in this work where the energy exchanges in the contact area are used to calculate the gas flow from the contact region, the effects of heat, enthalpy, arc power, mass flow rate, volume flow rate, and flow velocity in the contact area.

1.5 Summary of this book

This book presents valuable new insights into the arc root motion in the contact region. New results from pressure and spectral measurements are combined with optical data on arc root motion. This book consists of seven chapters:

The first chapter is an introduction that gives a general account of the arc root and the objectives of this work.

Chapter two presents a literature review of previous work in this field and gives information of arc concept. Relevant previous work in the arc phenomena field using the Flexible Test Apparatus (FTA) and Arc Imaging System (AIS) are summarised. This chapter also reviews literature relating to this new work.

Chapter three describes the Flexible Test Apparatus (FTA), the high speed Arc Imaging system (AIS), and the experimental equipment. The test procedures, experimental methodologies and techniques, are also explained. New pressure instruments and a spectrometer are used to observe the gas flow and gas composition effects on arc motion. Modifications to the software computer programmes and the experimental equipment are shown in this chapter.

Chapter four presents the experimental results of the arc root contact time, gas composition and gas flow pressure.

Chapter five presents the modelling results of the magnetic forces, gas dynamic forces, thermal energy and mass flow in the contact region.

Chapter six discusses the significance of the experimental results and effects of the experimental parameters. This discusses the influence of the arc chamber venting, contact material, arc chamber material, short circuit current level, supply polarity, contact opening velocity and the gap behind the moving contact. The modelling of gas flow and gas composition on arc root movement is also considered.

Chapter seven is the conclusion. This chapter shows how the results relate to the objectives.

1.6 Objectives

The objectives of the work programme are:

Design and modify the Flexible Test Apparatus (FTA) to allow the investigation of the gas flow and pressure behind the moving contact, at the fixed contact and behind the arc stack. The contact opening mechanism was required to operate at reduced velocity. Allow the contact opening velocity to be controlled at speed from 1 m/s to 10 m/s.

Modify the Arc imaging system (AIS) software programme to allow the acquisition of data of pressure measurement from the Flexible Test Apparatus (FTA) and analyse in the computer programme.

Use the Arc Imaging System (AIS) and the Flexible Test Apparatus (FTA) to investigate the anode and cathode root contact time in a miniature circuit breaker.

The main investigations involve the influence of arc chamber venting, short circuit current level, contact opening velocity, arc chamber material, supply polarity, contact material and the gap behind the moving contact.

Use the spectrometer to observe the arc spectrum and composition in the arc chamber when the contact is ignited. Using this technique also carry out an investigation into the effect of the contact material, contact opening velocity and arc chamber material.

Investigate the arc root commutation from the contact region at reduced contact opening velocity using the relationship of the magnetic and gas dynamic forces in the contact area.

CHAPTER 2

LITERATURE REVIEW

2.1 Introduction

In this chapter a study of the arc literature is divided into five parts. The first part is a basic arc theory placing emphasis on the arc root construction. The second part is the review of the previous work in the University of Southampton on the development techniques of the high speed imaging system (AIS). The experimental review of the relative work in this field is shown in section three. The composition of the arc is also described in this section. The modelling of thermal energy, magnetic and gas dynamic forces on the arc motion are examined in section four. The last section is the summary of the literature review.

2.2 Basic Arc Theory

The cathode contact provides the electrons to allow the arc to continue between the contacts. The cathode region can be described with a high electric field of 10^8-10^9 volts/meter. In general, the electron emission involves a combination of thermally enhanced field emission (T-F emission) and the effects of ion bombardment [4,5,6]. In the cathode fall region, about 90% of the current is carried by electron and 10% is carried by ions. The voltage drop in the cathode fall is approximately 15 volts. Cathode temperature is comparable to the boiling point of contact material. The high electron emission is produced by heat and enhanced field emission. The current density of the spot is about $10^3 - 10^6$ A/cm^2 [6].

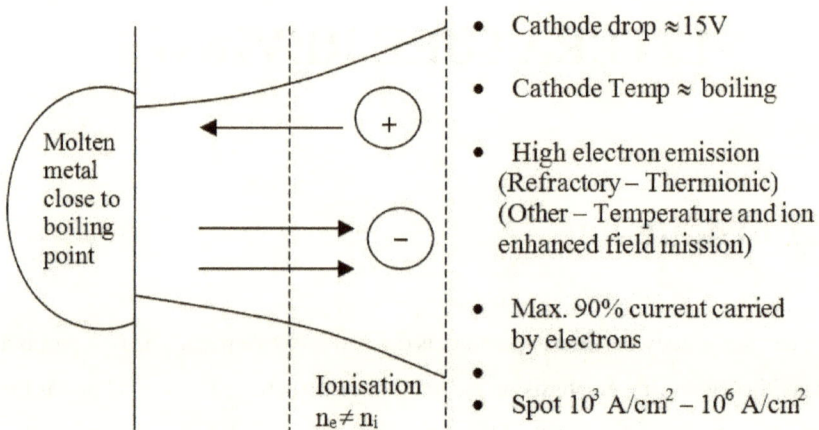

Molten metal close to boiling point

- Cathode drop ≈ 15V
- Cathode Temp $\approx$ boiling
- High electron emission (Refractory – Thermionic) (Other – Temperature and ion enhanced field mission)
- Max. 90% current carried by electrons
- Spot 10^3 A/cm^2 – 10^6 A/cm^2

Ionisation
$n_e \neq n_i$

Figure 2.1: Cathode fall [6]

The arc column has the characteristics of a plasma. The density of the electrons and ions are equal. In addition, the temperature of the electron and ions are equal to the gas temperature. More than 90% of the current in the arc column is carried by electrons [6].

The anode region serves to collect the electrons carrying the current from the arc column. The thermal boundary layer between the arc column and the anode surface is small. The electron density gradients are high so that electron diffusion flow exists. The anode fall voltages can be close to zero and as high as 15-20 volts. The anode fall temperature is about 200 degree C up to the boiling point of the contact material. The current is carried by electrons and anode spot current density is less than that of the cathode spot [6].

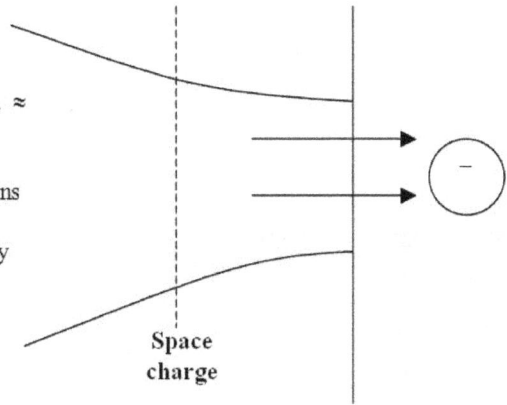

- Anode drop ≈ 15V

- Temperature range s from ≈ 200 °C to boiling

- Current carried by electrons

- Anode spot current density < cathode

Space charge

Figure 2.2: Anode fall [6]

An arc occurs at a typical voltage when contacts are opened to break an electrical circuit. The plasma of metallic ions is carrying an arc current within a space when the molten bridge explodes. The high temperature and electric field from positive ions are present at the cathode surface to maintain the emission of current between electrode surface and plasma.

Thus, the arc is ignited. The arc cannot exist if the arc current is lower than the minimum arc current. The value of this arc current is a characteristic of the contact material.

The minimum arc current is determined from the arc current which corresponds to extinction [24]. When the arc starts, the arc voltage steps up to the minimum arc voltage. This value can be evaluated from the summary of the cathode and anode fall voltage drops. On the other hand, the minimum arc voltage can also be calculated from the relationship of the work function of the contact metal and the ionisation potential of metal atoms.

During a short circuit fault an electric arc is drawn between opening contacts. In circuit breakers of current limiting design, the current through the conductors of the miniature circuit breakers generates a magnetic field in the arc chamber.

The electromagnetic and thermal forces of the arc are supplemented to force the arc away from the contact region along arc runners and directly into the arc stack. Here the arc is split into a number of series arcs. The resistance of the arc is then increased. This results in a high voltage which counteracts the supply voltage to limit the peak fault current.

In turn, the current is reduced. Therefore, the arc cannot be maintained. The resistance of the arc and the arc voltage can be varied by increasing the length of the arc, cooling the arc and splitting the arc into a number of series arcs. Most arc stacks use the metal plate type that splits the arc into a number of series arcs and also cools the plates by conduction.

2.3 Research work at the University of Southampton

This section reviews research at the University of Southampton as a preview to the work presented in this book.

Traditional methods using optical techniques to investigate the arc motion have recently been improved with the advent of solid-state high-speed imaging. An optical fibre array connected to an A-D (Analogue to Digital) circuit can allow image resolution of one million pictures per second of the arc covering the period of interest [9,10,11,12,13,14]. Techniques for recording arc motion in a miniature circuit breaker have improved significantly in recent years.

The high speed Arc Imaging System (AIS) used here has been used to record the light intensity of the optical data from an array of fibre optics located in the arc chamber. In addition, the AIS can provide a better time resolution when compared to both a streak camera and high-speed film techniques. The short circuit current is generated by the discharge of a capacitor discharge bank. The capacitors are charged from a rectified mains source [9,14,25,26].

The Arc Imaging System (AIS) was first built in 1990 by McBride and Weaver [9,10]. They started to use this system to study the motion of the electric arc with short circuit prospective peak current up to 3.4 kA. The arc was monitored by 22 optical fibres. These techniques permitted the visualisation of the arc ignition, arc immobility and the arc movement away from the contact region.

Afterwards, the array of fibre optics in the arc chamber was increased to 45 fibres. The motion of the arc movement from the moving contact was more clearly visible. The failure of the arc transfer from the moving contact was observed [12].

The first stage development of the test system was in the period between 1992-1994, the arc motion was investigated in combination with a pressure and conductance measurement, using 47 optical fibres in a commercial MCB. This time the system was used to monitor the short circuit current up to 16 kA [11,12,27].

The system allows identification of the arc root and position of the moving contact. The steel plates around the arc chamber generated effects on the magnetic field and the arc dynamics. The arc trajectories provided results of arc velocity from 0 – 500 m/s and a shock wave was formed as the arc started to move.

The second stage of the development of the test rig was in 1995-1998 [13,28,29]. A Flexible Test Apparatus (FTA) was built to simulate a short circuit current. The main structure was similar to the commercial circuit breakers (MCBs). The apparatus was adjusted to give control over the independent variables such as contact materials, chamber material, contact velocity, etc.

The solenoid was used to operate the contact mechanism to open the contacts with two divergent runner geometry. A linear contact mechanism with velocity of 1-3 m/s and a pivoting mechanism with higher velocity of 3-12 m/s were developed.

The AIS enabled the identification of the arc root period commutation from the fixed and moving contact. A centre of intensity method determined the position of the arc by an array of 75 optic fibres covering the arc chamber and allowed the study of the motion of both the arc roots along the runners.

The Taguchi method was used to investigate the arc behaviour in MCBs. The Taguchi analysis was used to show the best performance in terms of low let through energy and cathode root contact times.

The failure of the anode root to commute from the moving contact prevented the arc column displacing forwards of the cathode root. This experimental result was compared to a 3D (three dimensions) finite element model of electromagnetic field [14,30,31].

The third stage is presented here as the development and modification of the test system. The influence of the geometry of the arc chamber by closing the gap behind the moving contact, contact materials, arc chamber materials, are chamber venting, short circuit current and reduced contact opening velocity have been investigated.

The model of magnetic and gas dynamic effects in the contact region are used to explain the point at which the arc root commutates from the contact area. The discussion of the thermal enthalpy in the contact region assists in the analysis of the behaviour of the arc root motion. The study of gas flow, pressure and composition of the arc in the arc chamber are also considered.

2.4 Experimental research work

Numerous work [15,32,33,34,35] has studied the arc behaviour for opening contacts and observed that the arc does not move immediately from the ignition point, but remains at the contact area for a period of time. The arc root contact time is an indicator of the performance of circuit breakers. The arc root contact time depends on factors such as the contact opening speed [15,35], the contact material [15], and the arc chamber material [36,37].

2.4.1 Contact geometry and materials

Both contact geometry and contact material also have an effect on the arc immobility [25,37,38]. As the radius of curvature of a spherical contact was reduced, the arc immobilisation decreased. The arc root contact time on a Cu contact was found to be less than an Ag/C contact.

It was proposed that the arc on the Cu contact moved by the self magnetic field of the Cu conductor. This did not occur on the Ag/C contact since the self magnetic field was not of sufficient strength and any graphite content would reduce the arc running velocity.

When the arc commutation moved across a step or a gap, the cathode root ran along the edge and was immobile in the corner. The anode root could run along the surface and it would not remain in the corner [39] while McBride *et al* [34] showed that there was increased arc immobility at the step of the contact and the divergent corner of the arc runner.

Other factors that could have affected the experimental results include contact erosion, metal vapour, carbon contamination and wall degradation. For each current and arc chamber venting combination, every arcing operation changes the contact and arc chamber wall surfaces. The arc root mobility on the contact surface may also depend on surface topology [17].

2.4.2 Arc chamber material

The influence of insulating walls on the arc behaviour has been studied by many researchers [16,35,40,41,42]. The size and the shape of the arc chamber have a significant influence on the arc immobility. Ceramic walls affected the direction of the arc movement. Arc motion was effectively inhibited by gassing walls and the gassing walls impeded the arc on copper contacts. Ceramic walls did not obstruct arc movement.

The ablated surface influences the characteristics of plasma, causing increases in temperature, electron density and electric field, but reducing arc extinction time. The hydrogen molecule in the insulators could be regarded as favouring the break arc. The influence of hydrogen is reduced by the presence of metals of low ionisation potential which cause an increase in electric conductivity.

The pressure effect on the arc motion was investigated in [11,18,43,44]. Higher pressure induced an increase in gas flow and this high flow would result in a faster de-ionisation of the plasma.

The residual conduction occurrence was due to the presence of organic vapours coming from the ablation of the sidewall. The sidewall material may change the composition of the medium, thermodynamic properties and current distribution.

The pressure with the thermoplastic was higher than with the ceramic. A period of arc immobility in the contact region coincided with a period of constant pressure. The arc velocity may be limited by increasing flow speed in the re-circulation. The calculated properties of plasmas show that the hydrogen content affects the energy-carrying capacity of the plasmas through the transonic regions.

The effects of aromatic compounds on the contact resistance are well known [16,45,46,47,48]. Organic gas components are emitted from plastics, adhesives and sheaths and these cause an increase in the contact resistance. In an electrical life test, organic gas components create black powder deposits on the contact surface which increase the contact resistance.

The carbon contamination could reduce the effectiveness of the gassing materials as the carbon reacts with the oxygen and nitrogen. Interaction between the arc and Ag/C contact material in the surrounding atmosphere leaves CO and CN gas. Silver could decrease dielectric strength of a device. Condensed carbon could also have as much or more of an effect in reducing dielectric strength.

2.4.3 Arc chamber venting

Lindmayer, Shea and others [17,18] investigated the effects of the vent on the arc. The arc chamber venting area determines the pressure developed during an interruption and the flow of hot gas and metal vapour in the arc chamber. Furthermore, venting plays a key role in controlling the cooling rate of the arc.

The different vent sizes created an effect on plasma, with a large vent providing decreased plasma thermal temperature and a restricted vent size increased arc chamber pressure. When the arc chamber venting was limited, the arc's entry into the arc stack was delayed. The action of the current limiter of the circuit breaker was delayed and therefore, the let through energy value was increased.

2.4.4 Contact opening velocity

The optical fibre imaging and analysis system demonstrated a detailed and reliable study of the arc motion between opening contacts. A high contact velocity is an important but not essential parameter in providing arc root mobility [2].

In experimental studies of arc movement, where the contacts were opened by a spring-loaded solenoid with opening speed 4.1 m/s and a full open contact gap of 1.6 cm, the arc dwell time decreased with increasing contact separation speed [17].

Belbel [15] reported the arc root contact time was reduced to a minimum value when the contact opening velocity was above 6 m/s, this value being independent of contact material. The contact gap had little effect on the arc root contact time. Below 6 m/s, copper contacts were characterized by a consistently low immobility time regardless of the arc current.

Additionally for contact velocity below 2.2 m/s, the arc root contact time was reduced as the contact gap increased. The gas pressure increased when the contact gap was reduced. The experimental and analysis system is described in the following references [15,34,49].

Rieder [35] observed the minimum gap that the arc starts to move from the contact region. His experimental conditions were an arc current of 2 kAdc and a magnetic blast field 25mT/kA. The arc always commutated for a contact gap of 2 mm and it is independent of the contact material.

2.4.5 Short circuit current level

Widmann [39] presented the arc commutation delay decreases with increasing arc current. In his experiments, the values of arc current and the commutation delays on the anode and cathode root were measured. By the nature of the RLC circuit, the arcing times increased with increasing short circuit current levels.

Further increasing the short circuit arc current lead to arc development, raising the arc heat flux entering the contact, and increasing the gas pressure and rate of evaporation. A higher current produced a hotter and larger arc which then vaporized more arc chamber wall material creating greater amounts of gas, provided there was sufficient venting [17,50].

2.4.6 Arc spectrum

In the electric arc research field, a spectroscopic measurement is very important. A specific signal line was selected and converted to an electrical signal by a photo-multiplier. Spectrum intensity depended on the number and the transition probability of the luminescence atom and the molecule.

The spectrum of the same element, the difference of the excitation energy and the transition probability resulted in a difference of intensity. The mechanism of luminescence was the same for spectra. This conformed to the same element even if the wavelength was different [17,51].

The charge-coupled device (CCD) colour linear image sensor [52,53] combination of optical filters observed the spectrum of Ag I at wavelengths of 421 nm and 546 nm. The element of Ag was found at the wavelengths of 502.91 nm and 546.55 nm, carbon C at 613.4 nm, and O at 615.32 by the CCD image sensor [54]. Emission lines from copper were clearly shown at 515.3 nm and 521.8 nm, hydrogen at 486.3 nm and carbon at 516.5 nm by using a spectroscope.

The Ag I (atom) spectrum was detected not only in the metallic phase but also in the gaseous phase. In the metallic phase arc, the discharge was maintained in a metallic ion. This evaporated from the electrode metal immediately after breaking from the molten bridge.

In the gaseous phase, the molecule and ion of the surroundings gases contributed to the discharge. It was considered that the density of metallic vapour decreased due to the increase in gap between electrodes. Silver was depleted because of electron emissions from the contact.

For Ag/C contact material, where graphite particles were small, the carbon reacted with the oxygen and nitrogen of the surrounding atmosphere, creating a form of CO gas on the contact surface. The high percentage of the deposition of carbon from vapour species could increase the arc root contact time. The condensed carbon had much more of an effect in reducing the dielectric strength while silver on the surrounding arc chamber decreased the dielectric strength of a device [47,48].

2.5 Modelling

A model of three dimensional arc motions with current construction at the arc root region and Finite element magnetic field were derived [11,31,55,56]. There were two forces, electromagnetic and gas pressure forces, appearing at the closed contacts at the inrush of the short circuit current [57]. The steel in the sides of the arc chamber had a large effect on the magnetic field and the arc dynamics.

The contact gap reduced the magnetic field significantly. An inverse method was used to determine the evolution of a self-blown electric arc. The techniques of micro coils with induction measurement were used to define the speed of the anode and cathode arc root.

The magnetic blast due to conductors and electrodes varied rapidly in the electrode gap. The induction was very high close to the electrodes and decreased very quickly in the middle of this gap.

2.5.1 Magnetic forces

Paul modelled the magnetic forces in [11]. He considered the magnetic driving force in the contact region from the magnetic field from the conductors and steel plates in the side walls. The magnetic flux density (B) was 31 mT/kA for the conductors in the air with a gap of 10 mm. Thus, at an arc current 2000 A, the magnetic forces was about 1.24 N.

Three dimensions of arc motion between parallel arc runners were simulated to consider the interaction of magnetic forces, current flow, gas flow, heat conduction and radiation by using computation fluid dynamic (CFD) finite volume code [40]. The results of the simulation of the electric field distribution between two electrodes presented the arc shape and depended on the electric field strength distribution.

Although this simulation was a good presentation, a parallel arc chamber shape is not available in real miniature circuit breakers. To explain the arc root motion at the point that the arc root moves from the contact region at reduced contact velocity, the magnetic forces would be modelled as a function of the contact gap and the short circuit arc current at different contact opening velocity.

2.5.2 Gas dynamic forces

The effects of gas flow and wave processes on the arc motion were apparent in the arc chamber and had an important effect on the dynamics of the arc as the arc moved off from the contact region.

The gas flow patterns, magnetic and electric fields affected the shape of the core of the arc. The pressure distribution and the corresponding plasma formation were dependent on the cross sectional area of the arc chamber and outlet cross-section.

A wider chamber reduced the immobility time and enabled good pressure compensation inside the arc chamber. However, increased flow speeds in the re-circulation on the arc chamber, increased ablation from the sidewalls. Ablation from each type of arc chamber material produced different plasma compositions with different arc voltage and pressure characteristics [11,17,38,43,44].

2.5.3 Thermal energy

In some arc chambers, carbon and hydrocarbons along with the metal contact materials were redeposit back on the arc chamber walls and contacts. These deposits can affect subsequent gas evolution from the arc chamber surfaces and reduced arc chamber venting.

The arc current, arc chamber geometry, arc chamber material and pressure imposed by the arc chamber venting could limit the radius of the arc and the ablation rate of the arc chamber material.

In the three dimensional model of the effect of an external magnetic field and the presence of gassing materials [58,59], an ablation from the arc chamber material made of gassing material and the resulting mixing of hydrogen and air increased the arc voltage dramatically.

This was due to the increase in heat transfer around the arc body. The vapor pressure at the front of the evaporating cathode was a function of the air pressure and its value was estimated from the diffusion of the air pressure. The surface temperature of the evaporating cathode was close to the boiling state.

A large amount of the organic vapor which came from the vaporization of plastic side walls has been demonstrated in [60] and the high-pressure build up in the arc chamber confirms this interpretation.

John and David [61] used fiber optic arc motion superimposed on a CAD drawing to perform short circuit tests on digitized views. The acquired waveform showed that the pressure in the arc chamber was proportional to the arc power and not arc current.

The current density distribution of the arc lead to ohmic heating and magnetic forces which caused gas flow and energy transport within the plasma. This lead to a thermal and pressure distribution in the arc chamber [22]. At higher pressures, the effects of columbic interaction were significant. After vaporization, a plasma pressure could be extremely high and in arcs burning in the plastic arc chamber material. The ablation could cause very high pressures [62].

Gassing arc chamber material created a higher pressure which in turn lead to increased gas flow and arc voltage. High flow resulted in a faster deionisation of the plasma. The pressure in front and back of the arc after the arc ignition rose and was caused by the heating of the surrounding gas volume and the vaporization of the electrode and arc chamber material.

The pressure distribution and the plasma formation depended on the cross sectional area of the arc chambers. The related plasma chemical processes affecting the energy efficiency of the plasma chemical reactions were modelled. Increased voltage and high flow velocities provided intensive cooling, with an increase of electric field and decrease of gas temperature [17,18,63,64,65].

2.6 Summary

During the short circuit fault an electric arc is drawn between opening contacts and the current through the conductors of the miniature circuit breakers generates a magnetic field. The electromagnetic and thermal forces of the arc force the arc away from the contact region directly into the arc stack; the arc is then split into a number of series arcs. This results in a high voltage which counteracts the supply voltage to limit the peak fault current.

The Arc Imaging System (AIS) was used to record optical data of the arc motion at sample rates of 1 MHz. The Flexible Test Apparatus (FTA) was built to simulate a short circuit current. The short circuit current is generated by the discharge of a capacitor discharge bank. The apparatus was adjusted to give control over the independent variables.

As the arc commutation moved across a step, the cathode root was immobile in the corner while the anode root would not remain in the corner. This effect was clearly shown when a step was on the fixed contact. There are few investigations which observe the effect of the step on the moving contact on the arc root motion when the arc root moves from the contact region.

As the arc commutation moved across a gap, the arc root was immobile in the divergent corner of the arc runner. In typical arc chamber of commercial MCBs, there is a gap between the moving contact and the moving arc runner. The study of the effect of this gap on the arc root motion when the arc root leaves the moving contact could be beneficial in modifying the design of the commercial MCBs.

The arc chamber venting area determines the pressure developed during an interruption and the flow of hot gas and metal vapour in the arc chamber. Higher pressure induced an increase in gas flow and this higher flow resulted in a faster de-ionisation of the plasma. However, very little information is available today on the effects of the arc chamber venting observed under the conditions of reduced contact opening velocity.

The resolution of instrumentation in previous studies was limited to recording the arc motion in milliseconds, whist the period of the arc root motion when the arc root moves from the contact region was often shorter than 1 millisecond. With a high speed Arc Imaging System (AIS), this arc motion can be recorded, providing advantages in observing more details of the arc root motion.

The density of metallic vapour decreased due to the increase in gap between electrodes. At reduced contact opening velocities, the contact gap at the point the arc root moves from the contact region would be varied as the contact velocity. This would have effects on the gases composition in the arc chamber.

The observation of the arc spectrum in the arc chamber as the arc occurs could specify the chemical elements species which affect the arc root motion. In addition, this can be used to improve the arc control in low contact opening velocity.

The magnetic forces as a function of the contact gap and short circuit arc current at the point that the arc root moves from the contact region were modelled to explain the arc root motion at reduced contact velocity.

The gas dynamic modelling at reduced contact opening velocity can be used to explain the effect of the gas flow on the arc root motion events in the contact region when the arc root starts to move from the contact region.

The co-operation between the gas flow and the energy exchanges in the contact area and the heat deposited in contact area can be estimated and used to calculate the mass flow rate and volume flow rate in that area. The mass flow rate can be used to consider the total mass flow in the gap behind the moving contact when the arc root stays in the contact region, which affects the arc root motion at reduced contact opening velocity.

Previous studies showed that the arc root contact time was dependent on the contact opening velocity for contact opening velocities lower than 6 m/s. The arc root contact time was reduced to a minimum value when the contact opening velocity was above 6 m/s.

The arc root contact decreases as the contact opening velocity was increased for contact opening velocity from 2 m/s to 6 m/s. The arc always commutated from the contact region for a contact gap of 2 mm, it is independent of the contact material.

There are very little empirical works on reduced contact opening velocity. This is also lack of breadth in arc root motion at the point that the arc root moves from the contact region in low contact opening velocity circuit breakers.

CHAPTER 3

INSTRUMENTATION AND METHODOLOGY

3.1 Introduction

The information on the instrumentation and the methodology in this chapter is presented in twelve sections. Details of the Flexible Test Apparatus, high speed Arc Imaging System (AIS), pressure measurement and spectrometer are described in section one to five. The modification of the arc chamber gap behind the moving contact and contact velocity are provided in sections six and seven.

Sections eight and nine give information on experimental variable and fixed parameters. The experimental methodology is described in sections ten. The control and analysis computer programme is in sections eleven. The final section provides the information of the method for evaluating arc root motion.

Computational manipulation of optical data permits the identification of parameters describing the motion of both cathode and anode arc roots on the fixed and moving contacts. These instrument techniques were used to study the influence of arc chamber material, contact material, and contact opening speed on the arc root mobility during contact opening.

Figure 3.1 shows the schematic diagram of the arrangement of the Flexible Test Apparatus (FTA) and associated instrumentation. The arc chamber parameters were varied using the Flexible Test Apparatus (FTA) described previously [10,11,12]. The arc root mobility in the contact region has been investigated individually on the cathode and anode arc root movement.

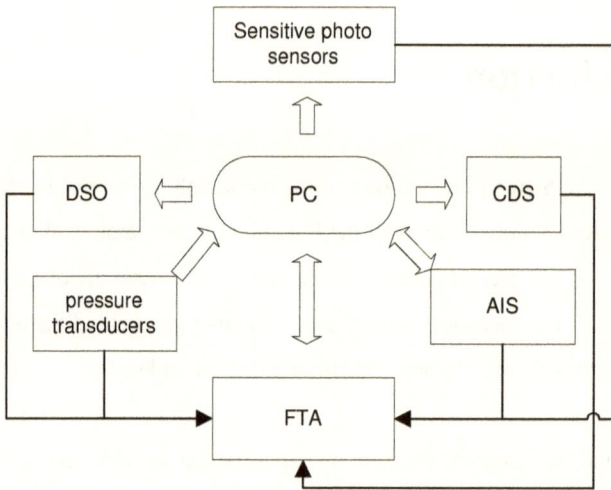

Figure 3.1: Schematic diagram showing the arrangement of the Flexible Test Apparatus(FTA) and associated instrumentation

3.2 Flexible Test Apparatus (FTA)

The Flexible Test Apparatus (FTA) was used to simulate the operation of a miniature circuit breaker when a short circuit current occurs. The highlights of the FTA are the independent configuration of the contact mechanism, repeatable contact action, variable contact material, variable arc chamber geometry, variable arc chamber and variable arc chamber vent configuration.

The overall details of the FTA are shown in Figure 3.2-3.4. The FTA is bolted to an earthed aluminium plate. The structure of FTA is similar to commercial MCBs. The main components consist of solenoid, arc chamber, moving contact, fixed contact, arc runner, arc chamber vent and arc stack. An array optical fibre is placed on top of the arc chamber.

The arc chambers are machined from Tufnol material. Arc chamber sidewalls are made from Macor machineable ceramic. The top plate is a transparent clear view to allow the arc motion to be recorded by the AIS. The bottom plate is changeable to study the effect of arc chamber material.

The moving contact mechanism is opened by the impact of a hammer connected to the solenoid. When the hammer passes the optical sensor, the trigger signal is sent to count down at the control computer to start the short circuit current fault and data acquisition.

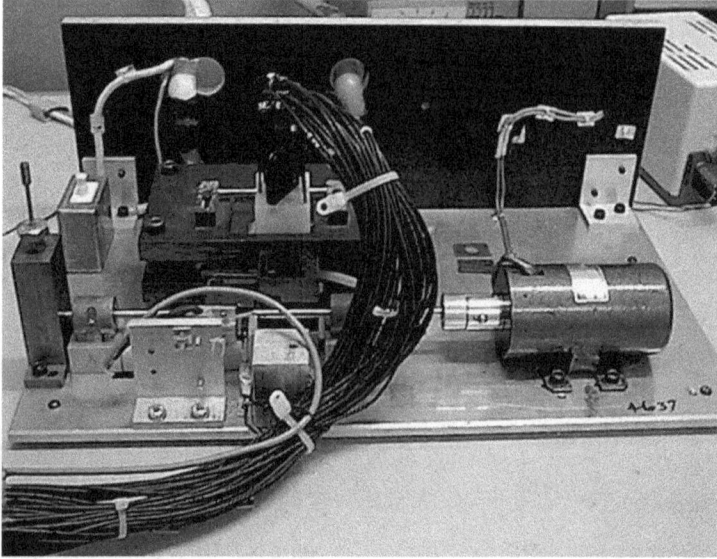

Figure 3.2: Flexible Test Apparatus(FTA)

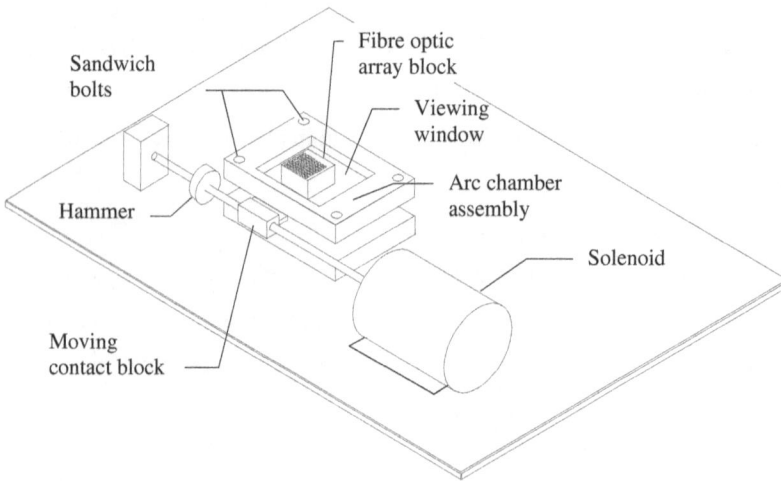

Figure 3.3: Structure of the Flexible Test Apparatus (FTA) (unscale)[2]

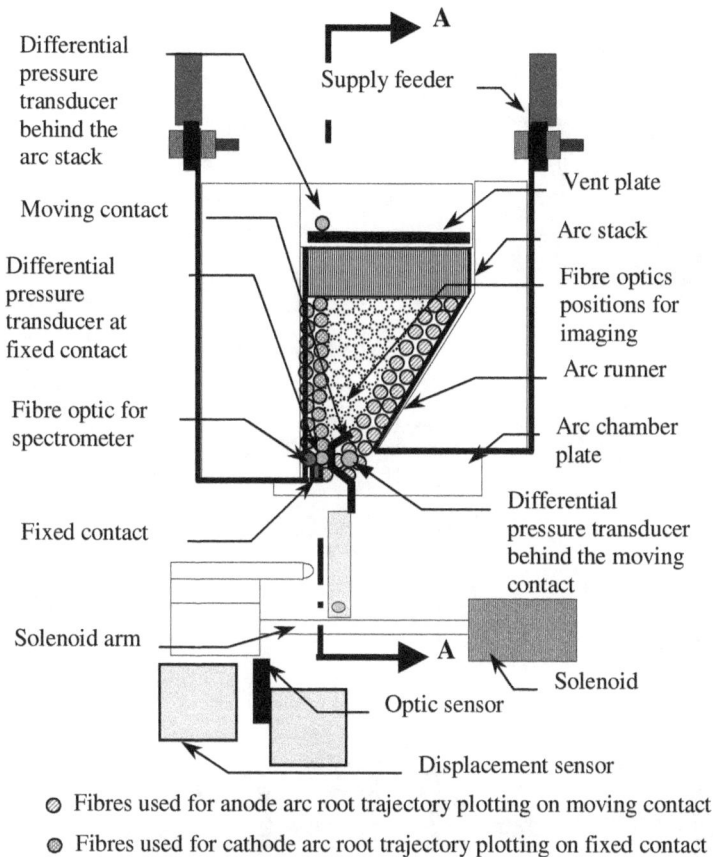

Figure 3.4: Flexible Test Apparatus component

The dimension of the arc chamber and position of the fibre optic is shown in Figure 4.2, Chapter 4. A cross section of the arc chamber in the Flexible Test Apparatus (FTA) and the arc chamber are shown in Figure 3.5

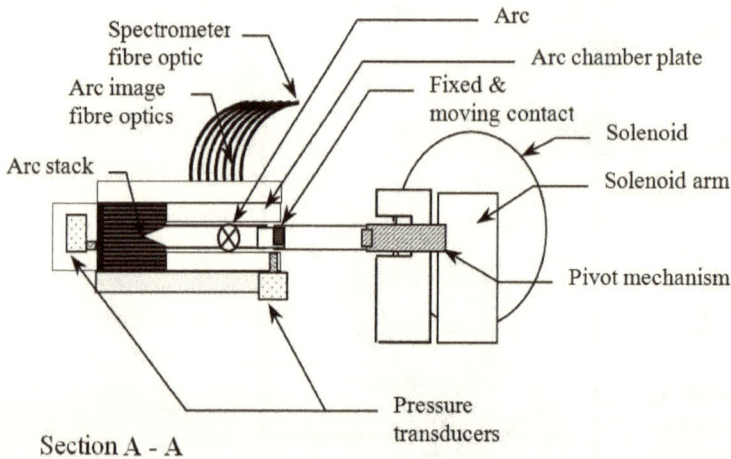

Section A - A

(A) Cross section of the Flexible Test Apparatus (FTA)

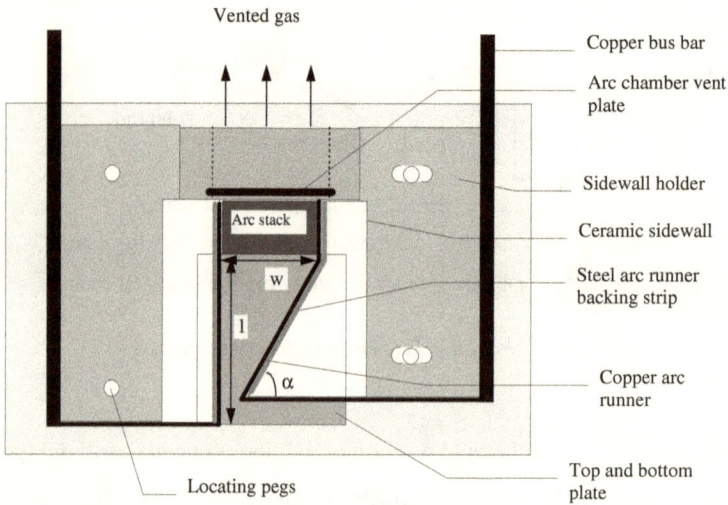

Angle of divergence α=60°. Chamber length l = 45 mm, Chamber width = 29mm, Chamber depth d = 6mm.

(B) Arc chamber geometry in the Flexible Test Apparatus (FTA)

Figure 3.5: Cross section and arc chamber geometry in the FTA [2]

Commercial devices are variable. Many of them use ceramic side plates in the arc chamber. The ceramic is a good insulator, but also acts as a heat sink. Ceramics are used in MCBs , but may be more prevalent to larger circuit breakers such as moulded case breakers. The ceramic arc chamber in the Flexible Test Apparatus (FTA) would be quite similar to this.

Some devices use plastic parts for the side walls which would have different thermal properties. The effects of different materials combinations are part of the work in this book. This is a substantial body of work into different materials and their effect on the arc control. In away, this whole area of work needs to be re-applied to the optimisation of a low contact velocity device.

3.3 High Speed Arc Imaging System (AIS)

The Arc imaging System (AIS) is employed to record the optical data from the arc chamber in the Flexible Test Apparatus (FTA). Each optical fibre has a defined position relative to the interior components of the miniature circuit breaker. Polymer fibre with 1 mm. core diameter surrounded by a 0.5 mm sheath was used. The attenuation of polymer is 200 dB/km at 665 nm [66].

A dimensional array of fibre optics is placed over the arc chamber of the FTA, as shown in Figures 3.4. Each fibre optic records the light level at a specific location within the arc chamber. The arrangement of the fibres within the fibre optic array was customised for each arc chamber geometry used in the investigations. The location of the fibres used for each arc chamber is shown in Figure 3.4 and Figure 4.2, Chapter 4.

The fibres are hexagonally close packed with centres 3 mm apart. Each optic fibre in the array is recessed in the fibre optic array in a round hole as shown in figure 3.6.

The radius of view at the centre of the arc chamber depth is calculated by $r = t (0.5 + d/a)$, where t is the fibre diameter (1mm), d is the depth to the viewing plane (15 mm) and a is the depth of the fibre recess (25mm). This gives a viewing radius of 1.1 mm for each fibre optic [2].

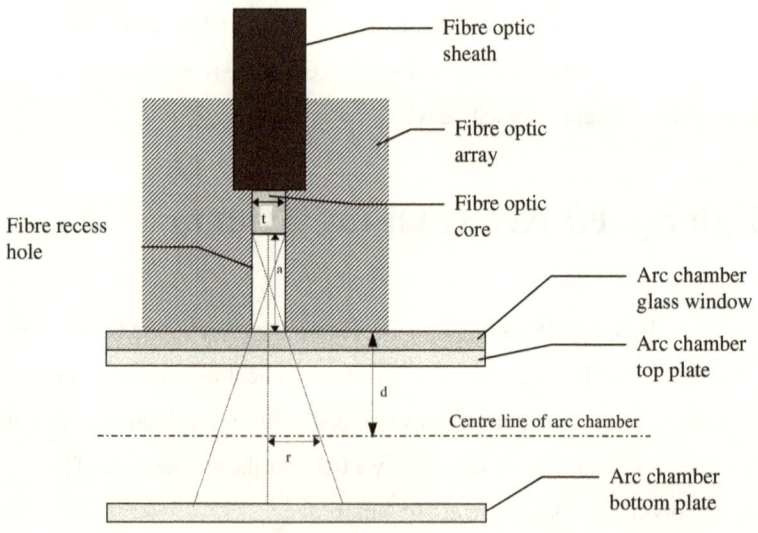

Figure 3.6: Fibre optics acting as a pinhole camera [2]

The light from each fibre is converted to an electrical signal by the optical circuit as shown in Figure 3.7. A photodiode is used to convert the light transmitted through the optical fibre into an electronic signal. Then the analogue signals are multiplexed into a group of eight and then converted to digital format by A/D converter as shown in Figure 3.8.

Data is retained in the random access memory (RAM) of the AIS. A digital I/O card in the personal computer is then used to transfer the data [9,10,11,66]. The schematic diagram of the AIS is shown in Figure 3.9.

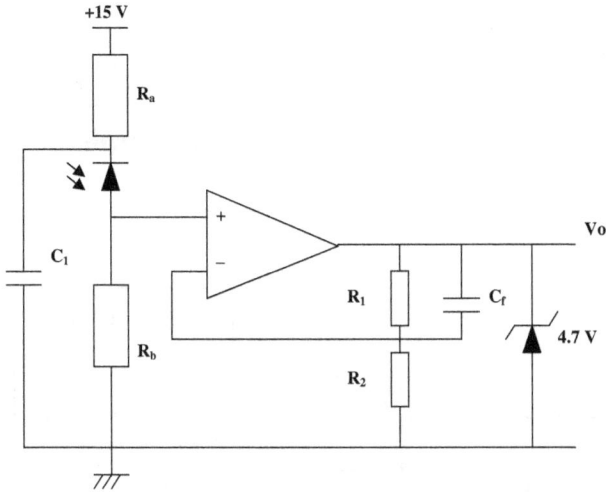

Figure 3.7: Photo-detector and amplifier circuit [9]

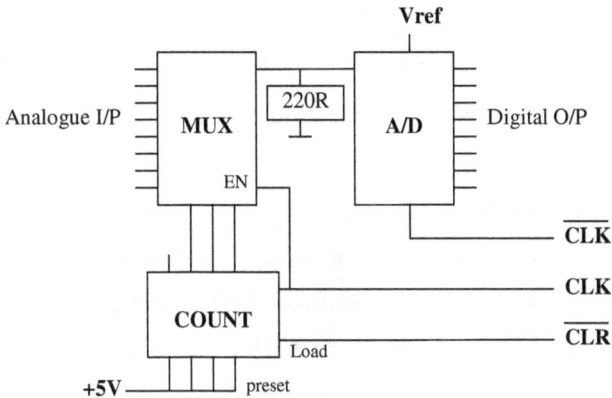

Figure 3.8: Analogue multiplexer and A/D converter [9]

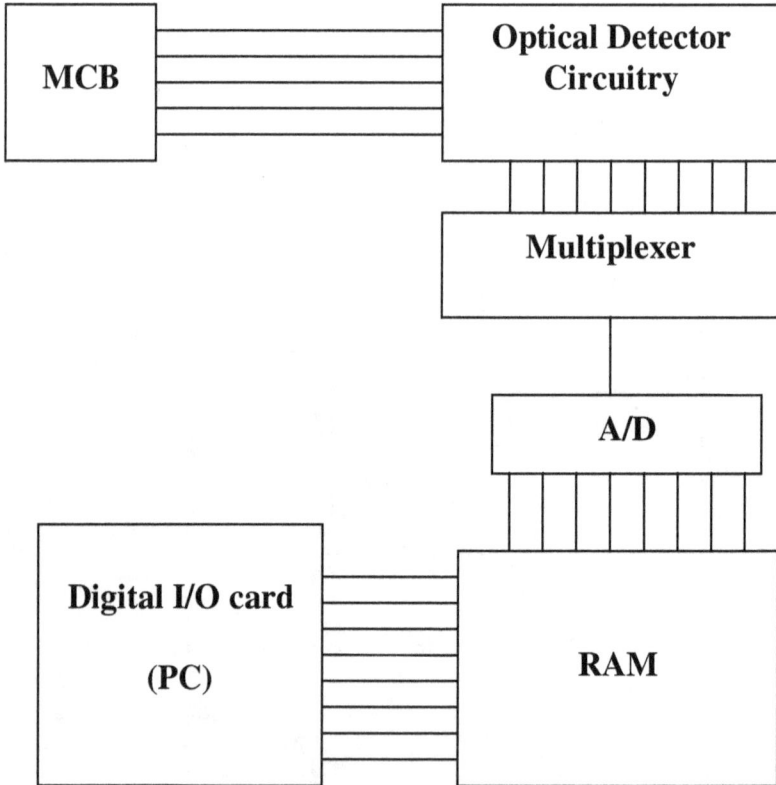

Figure 3.9: Schematic diagram of data recording in the Arc Imaging System [9]

The AIS samples light levels from each fibre optic at a rate of a million samples per second, and it is capable to record the arc movement for 8 ms. Each complete data acquisition card can handle 15 fibre optics cables. Each card features 15 photo-transistors, two 8 way multiplexers, two 6 bit 8 MHz Flash A-D converters, 32K of Ram and also the control & timing circuit [9,10,11,66].

3.4 Pressure measurement instruments

Two pressure transducers are used to monitor the gas pressure in the arc chamber. The SX series of piezo resistive pressure sensor functions as a wheatstone bridge on a silicon chip with a response times of 0.1 ms.

The pressure transducer gives a voltage output directly proportional to the applied pressure with sensitivity 0.75 mV/Psi. Long term stability is 0.1% and repeatability is 0.5% (Maximum difference in output at any pressure with the operating pressure range and temperature within O degree C to +70 degree C). Resistors are ion implanted into the silicon. The cavity etched on the reverse to create a thin silicon diaphragm [67].

The instrument amplifier AD621 was used to amplify the signal. The pressure transducer instrument was calibrated against a gauge pressure standard. Differential pressure measured relative to atmospheric.

The pressure transducers were installed in the fixed contact region, in the gap behind the moving contact, and behind the arc stack. The locations of pressure transducers as installed in the FTA are shown in Figure 3.4. The pressure transducer connection and pressure transducer circuits are shown in figure 3.10 and 3.11.

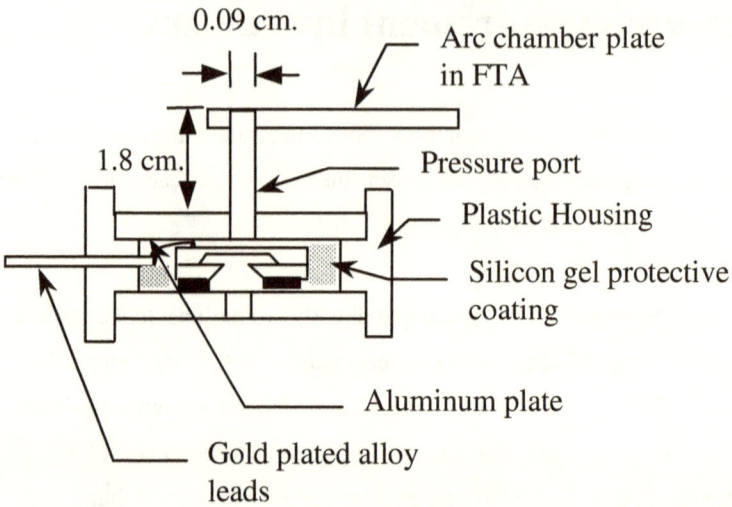

0.09 cm.

1.8 cm.

Arc chamber plate in FTA

Pressure port

Plastic Housing

Silicon gel protective coating

Aluminum plate

Gold plated alloy leads

Figure 3.10: Installation and connection of pressure transducer

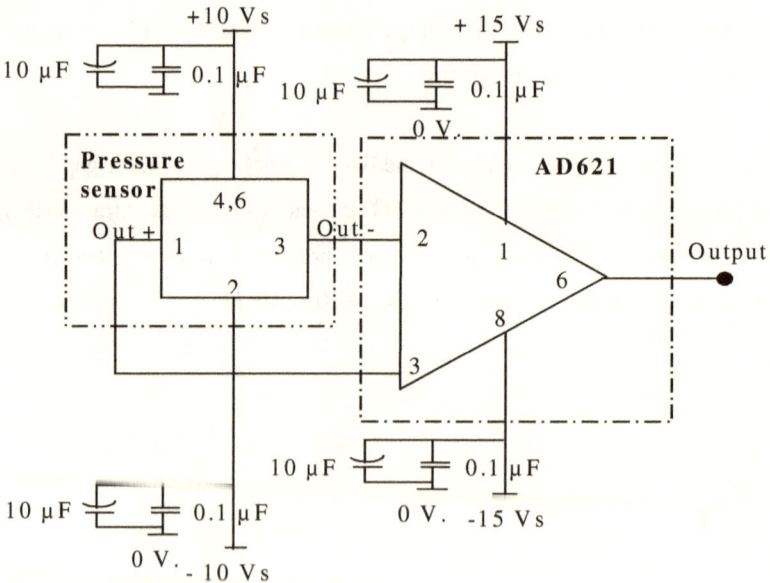

Figure 3.11: Pressure measurement circuit

3.4.1 Accuracy of pressure measurement

The accuracy of the pressure measurement was tested as details as follows:

Vibration effect: This is concerned about the vibration on the pressure transducer transmitted through the Flexible Test Apparatus (FTA). This was shown to be minimal.

Environment effect: This is concerned on the effects of the environment (air inside the tube with 1.8 cm. Length). This tube is connected from a pressure transducer to the arc chamber bottom plate in such a way to minimise environmental effects.

3.4.2 Electrical Noise

Power supplies: Each pressure transducer is connected to a difference power supply and separated amplifier circuit.

Cables: Three power cables from +Vs, GND and -Vs are twisted together to protect cross talk and noise.

Length of cable: the length of cables is limited as short as possible and close to the pressure transducer circuit.

3.4.3 Ground system

All of ground connectors for all experiments were connect together to the earth aluminium plate and aluminium boxes of the amplifier circuit.

3.5 Spectrometer

The optic fibre spectrometer was installed to measure the spectral data during arcing. A computer (PC) based S2000 plug in miniature optical fibre spectrometer as shown in Figure 3.12 was used to monitor the gas composition in the arc chamber.

The S2000 miniature fibre spectrometer accepts the light energy transmitted through an optical fibre and disperses it via a fixed grating across a 2048 element linear CCD array detector. This is responsive from 200-1100 nm. A wavelength of any spectral line has a maximum error of +/- 0.68 nm, sufficient for the rather broad spectral lines studied here.

Spectra were observed by means of a gated optical multi-channel analyser. The light is then collected and transmitted through the read fibre to the spectrometer. The spectrometer measures the amount of light. The ADC 1000 A/D converter transforms the analogue data into digital information [67].

The performance specifications of the S2000 miniature optical fibre spectrometer are the configuration featured a 600 lines/mm grating and 1000 nm optical fibre for signal collection with 14 grating UV through short wave NIR. The focal length is 42 mm for input and 68 mm for output.

Figure 3.12: S2000 plug in miniature optical fibre spectrometer

The optical resolution is 0.3 to 10.0 nm depending on grating and size of entrance aperture. Dynamic range is 2×10^8 for system and 2000:1 for a single scan. The sensitivity is estimated about 86 photons/count; 2.9×10^{-17} J/count, 2.9×10^{-17} W/count for 1 sec integration time. The integration time can be set from 3 ms to 60 sec with1 MHz A/D card. The fibre optic connector is SMA 905 to single strand optical fibre.

The integration time of the spectrometer is analogous to the shutter speed of a camera. The higher the value specified for the integration time, the longer the detector looks at the incoming data. Spectra can be exported into a comma-delimited ASCII file and saved as a binary file.

3.5.1 Spectrometer installation

Details of the optic fibre spectrometer installation in the Flexible Test Apparatus (FTA) are shown in Figure 3.13. The optical fibre of the spectrometer is placed cover on top of the arc chamber. The location can be selected from the fixed contact to the moving contact, along the fixed and moving arc runner and along the line in front of the arc stack.

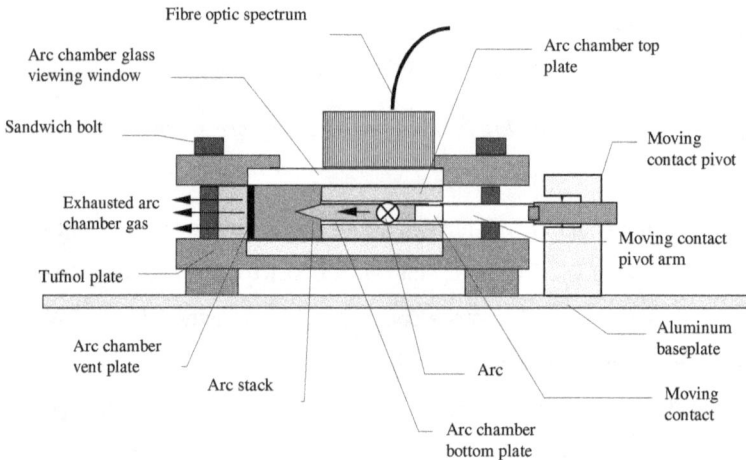

Figure 3.13: Installation of the fibre optic for measurement spectrum in the Flexible Test Apparatus (FTA)

Typical data from the miniature optical fibre spectrometer is shown in Figure 3.14. The commercial package software programme OOOB for the miniature optical fibre spectrometer is used to detect the signal of the arc spectrum. It presents the wavelength (nm) in x-axis and the relative intensity in the Y-axis of each spectrum line.

This programme has the ability to perform spectroscopic measurements such as absorbance, reflectance and emission, control all system parameters, collect data from spectrometer channels simultaneously and display the results in a single spectral window.

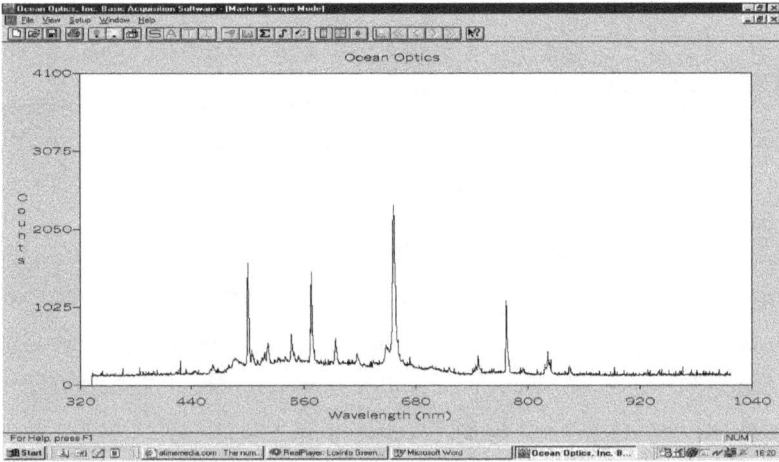

Figure 3.14: Capture of the arc spectrum

3.5.2 Accuracy of the spectrometer

To evaluate the accuracy of the output spectrum wavelength from the spectrometer, four tests with standard colour lamps have been examined the output of the spectra wavelength. The standard lamps are red, white, green and yellow. If the output data is significant, it would be expected to show the spectra wavelength near the standard value. Table 3.1 presents the output test results with standard lamps.

Standard colour lamps	Standard wavelength (nm)	Measurement wavelength (nm)		
		Lower	Peak	Upper
Red	620-700	620	650	725
White	515-700	475	580	725
Green	490-580	430	545	600
Yellow	520-700	525	590	730

Table 3.1 Test results from the standard colour lamps

The output results of the measurement spectra wavelength show that the peak of the spectrums are in the range of the standard values with a tolerance at the lower wavelength 0 nm, -40 nm, -60 nm, +5 nm and at the upper wavelength +25 nm, +25 nm, +20 nm, +30 nm for red, white, green and yellow colour lamps, respectively. A wavelength of spectral line has a maximum error of +/- 5%.

3.6 Contact velocity

The opening contact mechanism was modified to reduced contact speed from 10 m/s down to 1 m/s. There are two important parts in this system. The first part is the instrument to measure the speed of contact and the second part is the instrument to control the speed.

The instrumentation used to measure the contact velocity is the displacement sensor which is connected to the solenoid. The method used to control the contact velocity is to control the supply voltage to the motor of the solenoid.

3.6.1 Displacement sensors

The pivot mechanism moves in a curve with a very high speed, it is very difficult to measure speed directly from this system. It is not very difficult to measure the velocity of the solenoid movement and calculate the speed of the pivot mechanism and contact. The displacement transducer uses a filter pattern together with an intensity sensitive photo sensors. Details of the displacement transducer control circuit and differential op-amp circuits are shown in Figure 3.15 –3.16.

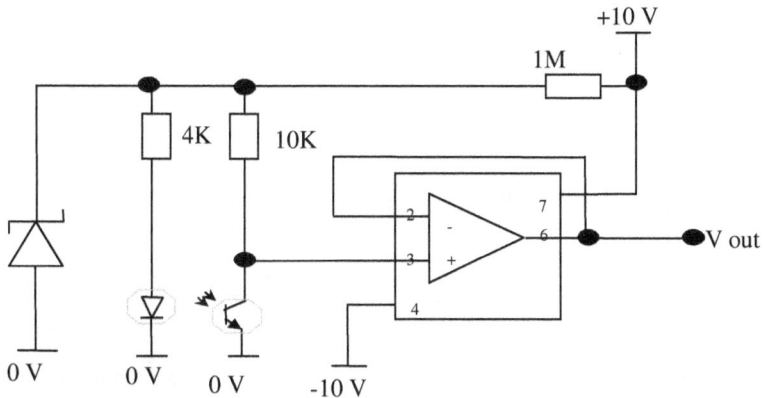

Figure 3.15: Control diagram of displacement transducer

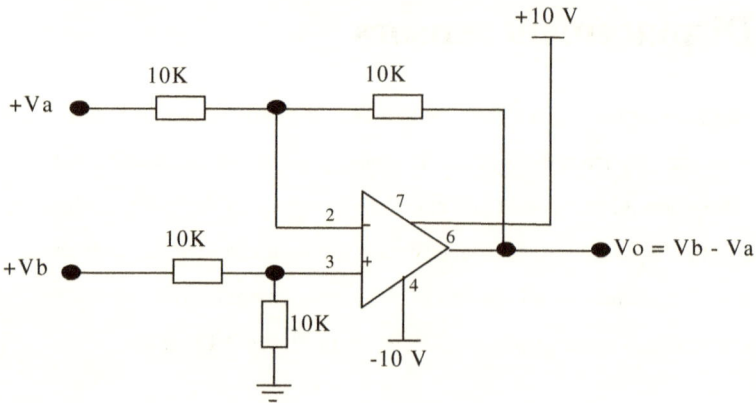

Figure 3.16: Detail of the differential circuit for output signal of displacement transducer

3.6.2 Contact speed control

The velocity of the solenoid movement can be controlled by reduced power supply voltage. The voltage across the armature and the strength of the magnetic field are the main factors which affect the speed of the solenoid [68]. This can be controlled by many methods i.e. shunt field, armature resistance, variable voltage and multi-voltage. In this work, the armature resistance method is used to control speed.

The wiring diagram to control contact opening velocity by inserting a variable resistor in series with a supply solenoid is shown in Figure 3.17-3.18 to reduce the speed of solenoid.

The supply voltage of the solenoid is decreased due to the voltage drop across the series resistance. The contact velocity can be regulated between 10 m/s and 1 m/s with average tolerance +/- 0.02 m/s.

The contact opening velocity is measured as the displacement of the solenoid arm while moving pass the optic sensor. For repeatability of the contact opening velocity from 10 m/s to 1 m/s, more than 15-20 experiments have been tested. The accuracy of the contact opening velocity of 10 m/s is approximately +/- 0.01 m/s and +/- 0.02 m/s for contact opening velocity at 5.5 m/s, 4.0 m/s and 1 m/s.

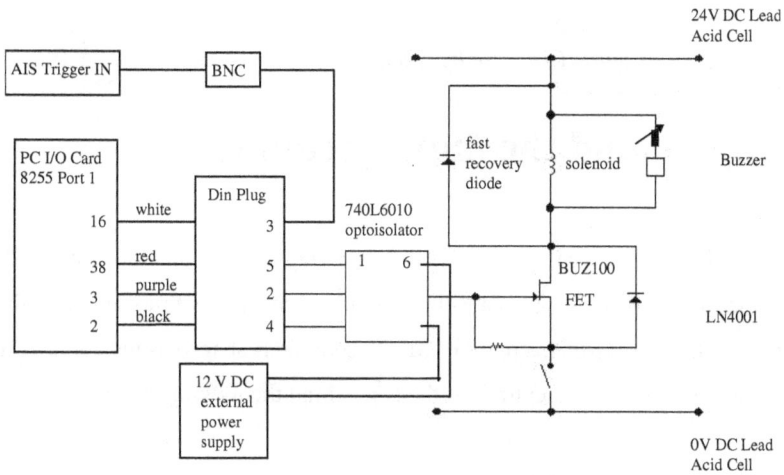

Figure 3.17: Existing solenoid control circuit for FTA [2]

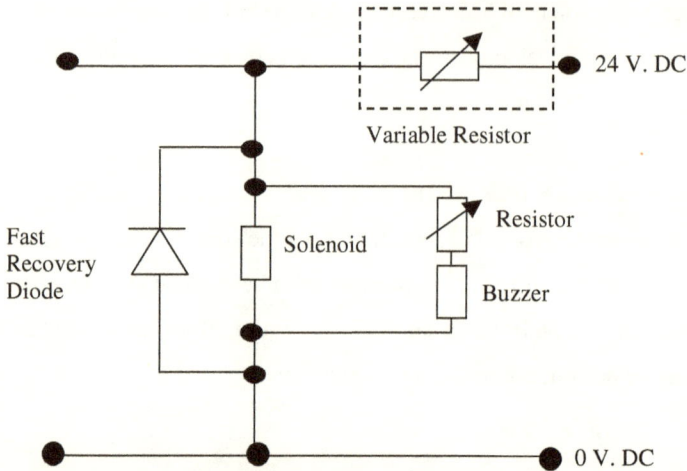

Figure 3.18: Modify solenoid control velocity

3.7 Gap behind the moving contact

In commercial MCBs, there is a gap behind the moving contact. It is thought that the gas flow escapes through this gap causing the delay of the arc root movement in the moving contact. To study this, a piece of Macro ceramic is shaped as the curve of the geometry of the moving contact to limit the gap behind the moving contact.

This ceramic is placed behind the moving contact so that when the moving contact is fully opened, there is minimal gap. Figure 3.19 shows a dimension and installation of the closed gap ceramic.

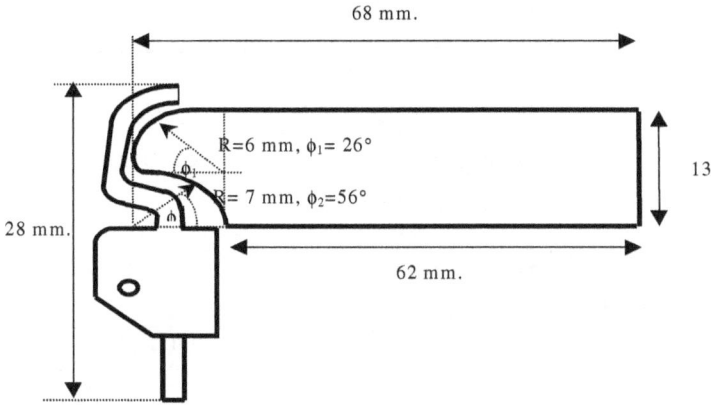

Figure 3.19: Dimension and installation of closed gap ceramic

3.8 Experimental variable parameters

There are seven variable parameters used to investigate the effects on the mobility of the arc root from the contact region. They are contact material, arc chamber vent, peak short circuit current level, contact velocity, arc chamber material, the gap behind the moving contact and supply polarity.

The details of the variable parameters to be investigated are shown below and shown in Figure 3.20.

Contact opening velocity	10 m/s, 5.5 m/s, 4 m/s, 1 m/s
Supply polarity	- Anode on the moving contact and cathode on the fixed contact - Anode on the fixed contact and cathode on the moving contact
Peak short circuit current level	2000 A, 500 A, 1400 A.
Contact material	Ag/C (95/5) step on the moving and fixed contact Cu punch on the fixed contact Ag/C (95/5) flat on the fixed contact
Arc chamber venting	Opened, Choked, Closed
Arc chamber material	Ceramic, Plastic polycarbonate
Gap behind the moving contact	Opened, Closed

Table 3.2: Experimental materials

Component parts in the FTA

Ag/C flat on the fixed contact

Cu punch on the fixed contact

Ag/C step o the fixed contact

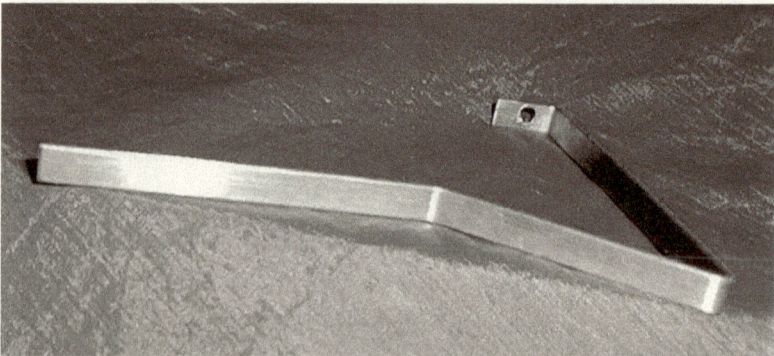

Moving contact arc runner (typical)

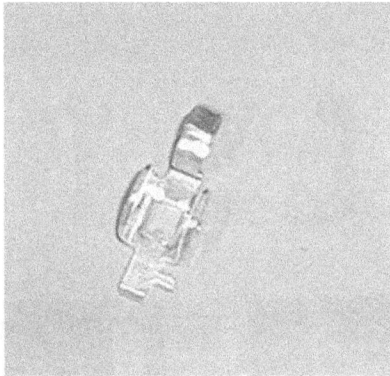

Ag/C step on moving contact

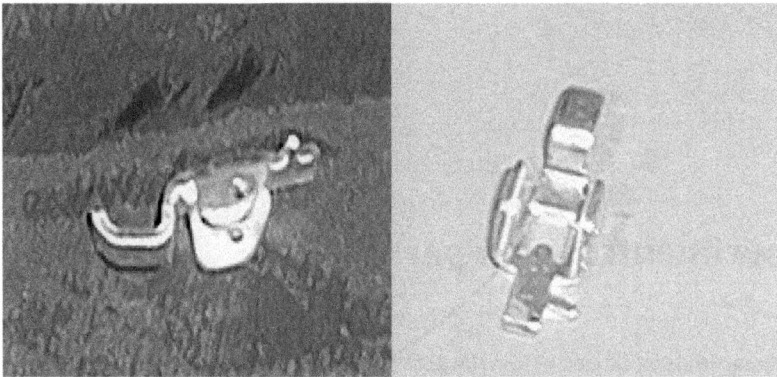

Moving contact (typical

Polycarbonate and ceramic arc chamber

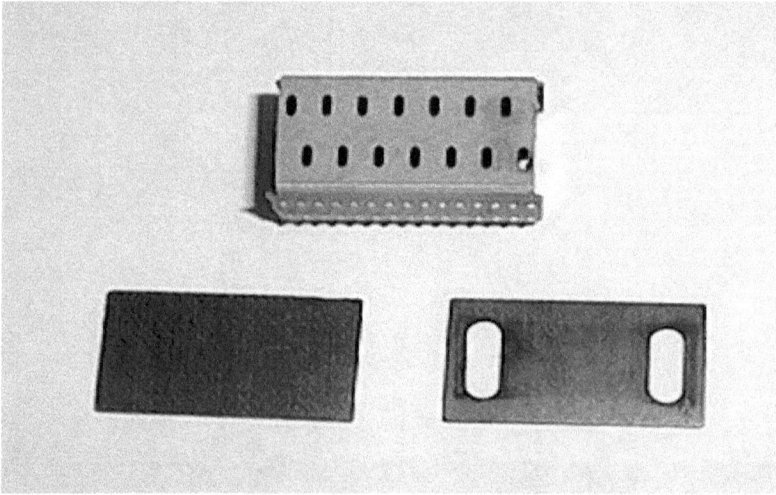

Vent: Opened, closed, choked

Figure 3.20: Experimental materials

3.9 Experimental fixed parameters

The experimental study is carried out on the arc root motion when the arc is drawn from the fixed and moving contacts until the arc reaches the arc stack. The arc displacement, arc current, arc voltage, pressure in the arc chamber and spectra of the arc are presented for each experiment.

The behaviour of the arc is then analysed from these plots. Experimental constants for these experiments are defined in Table 3.3.

Experimental Factor	Fixed value
Discharge system Inductance	224 μF
Discharge system Capacitance	47.4 mF
Final contact gap	6 mm.
Contact mechanism	Pivoting mechanism
Moving contact	Silver plated copper
Contact opening delay (t_{cod})	500 μs

Table 3.3 Experimental constants

3.10 Software computer programmes

A personal computer (PC) is used to control the sequence of the associated instruments. The computer (PC) is fitted with twin 16 channel input/output cards (I/O A and I/O B) with an onboard timer. The first I/O card is applied to control the AIS.

The second I/O card is utilised to control the capacitor discharge bank, Flexible Test Apparatus (FTA) and also to trigger the Digital Storage Oscilloscope (DSO). There are three computer programs, all have been written in Qbasic computer programme.

These programmes are to control the operation of the experimental equipment, analysing the experimental results and make a series of the arc images movies to observe the movement of the arc root from contact region into the arc chamber. Description of the programmes as follows:

Flexgav4.bas: This program is installed to run the test sequences operation. This program is used to control the sequences of the Flexible Test Apparatus (FTA), Arc Imaging System (AIS), the capacitor discharge system (CDS) and the digital storage oscilloscope (DSO).

The software computer program Flexgav4.bas starts to run, the delay contact time is entered as required, the capacitor bank is then charged to the desired voltage, the apparatus operate sequentially.

The solenoid is fired and the test data is recorded in the RAM of the Arc Imaging System (AIS) as previously described [9,10,11,12,66]. Simultaneously the short circuit current and the short circuit voltage are displayed on the digital storage oscilloscope (DSO).

The experimental result is transferred from the memory of the digital storage oscilloscope (DSO) and from the Arc Imaging System (AIS) to the hard disk in the personal computer (PC).

When this program is initiated, the Digital Storage Oscilloscope (DSO) will reset its system. A symbol READY status for recording a new test data acquisition is shown. A delay contact time (t_{cod}) is entered into the PC. After that the apparatus will start to charge the capacitor bank until the required voltage is reached.

Arcimage.bas: This program is able to present a series of images of the arc motion from the contact region until the arc reaches at the arc stack as a movie. It also replays the arc image contour movies.

Rootplot.bas: This program is used to analyse the experimental results. It shows cathode root contact time and anode root contact time separately with a facility to plot the arc root displacement. The arc power, arc voltage, arc current, etc also plot together with the arc root displacement.

3.11 Experimental methodology

3.11.1 Flexible Test Apparatus (FTA) set up

All surfaces are cleared after the arc chamber wall is removed from the test rig. The new copper arc runner and the steel backing plate are cleaned before they are fixed into the arc chamber. The moving contact is cleaned before it is soldered to connect with the copper braid. After that, the moving contact is fastened into the moving contact block assembly.

The copper braid is fixed with the moving contact runner into the arc chamber wall. The arc stack is put behind the arc chamber. A quartz glass is inserted above the arc chamber until it touches closely to the arc stack.

The top Tufnol plate with the fibre optic array is placed over the top of the arc chamber and fixed with bolts. Now the Flexible Test Apparatus (FTA) is ready to start a new experiment.

However, it is important to make sure that the current flow continuously between the fixed contact and the moving contact. The experimental results are presented here as an investigation of the arc root contact time, the average and the standard deviation of the anode and cathode root contact time on the fixed contact and on the moving contact.

After all of materials are prepared in the Flexible Test Apparatus (FTA). The solenoid's arm, which directly connects to the hammer, is manually moved to the maximum displacement. Next, The Digital Storage Oscilloscope (DSO), the Arc Imaging System (AIS), the control computer and the solenoid power supply are switched on.

3.11.2 Variable factors

To investigate the influence of Ag/C contact materials on the arc root motion from the contact region, a piece of Ag/C is welded on the fixed and the moving contact when the anode and cathode power supply are set in the fixed and moving contacts.

To inspect the influence of the short circuit current level on the arc root commutation from the contact region, three levels of the short circuit current are expected: 500 A, 1400 A and 2000 A. The Capacitor Discharge System (CDS) is charged up with capacitor 47.4 mF, inductor 224 μH and resistance 30 milliohm [2].

To investigation the influence of the arc chamber materials on the arc root moving off from the contact region, a variety of arc chambers are created using interchangeable components. The top and bottom plates are plain flat rectangles of material. The plates are squeezed firmly into the assembly.

The top pate is transparent to allow the high Speed Arc Imaging System (AIS) to record the arc motion in the arc chamber. The top plate is used the quartz glass and the bottom plate is used the polycarbonate or ceramic.

There are three vent plate configurations used in these experiments as shown in Figure 3.19 to investigate the influence of the arc chamber venting on the motion of the arc root commutes from the contact region. They are opened plate which has a vent area 40 mm^2, choked plate (vent area 15%) and closed plate which there is no vent area. The venting of the arc chamber is regulated by using a rectangular aperture with slot to hold a vent plate.

In order to observe whether the gap behind the moving contact affects the arc root moving off from the contact region, a piece of Macor ceramic is machined to match the shape of the moving contact to limit the gap behind the moving contact. The experimental results when the gap behind the moving contact is closed and opened are then compared.

In all cases the methodology used is well established and involves using new materials in the test after 10 consecutive short circuit tests. There are 10-30 tests per condition repeated. The experimental results are presented with the error band of ±1 standard deviation.

3.11.3 Spectrometer set up

To investigate the chemical elements of the arc in the arc chamber, the optic fibre spectrometer is used to observe the spectrum of the arc. The wavelength of the spectrums provides information of the chemical elements which mix in the hot gases [69].

Start the computer and select the OOIBase icon to run the programme. Enter user name and serial number to register in the operator and serial number dialog box. Then setting the hardware configure by choose the type of spectrometer and A/D converter.

Finally, select spectrum/configure data acquisition to the configure data acquisition dialog box. Set the integration time and specify the external trigger mode. The setting values are shown in the operating manual and user guide of the S2000 Miniature fibre optic spectrometers [67]. The monitor shows window of the real time of the spectrum capture. Now the spectrometer operating software is ready to capture the spectrum.

The optical fibre of spectrometer is installed on the cover of the arc chamber in the Flexible Test Apparatus (FTA). Press the icon CAPTURE on the top of window at the same time as press the SPACEBAR of the control computer to energise the solenoid.

The waveforms of the arc spectrum are shown in the window. The output data can be saved to disk as ASCII files or print the spectra or graphical data into Microsoft excel or Word.

3.11.4 Pressure transducer set up

To investigate the influence of the pressure in the arc chamber on the arc root movement, the pressure transducers are installed in three locations inside the arc chamber.

A two layer thermoplastic and metal tube, is connected into the pressure transducers. The end of these tubes are glued together to the base of pressure sensor to protect gas leaking. The pressure sensor is then connected into the pressure measurement circuit, amplifier circuit and power supply.

The output probe is connected into the digital storage oscilloscope (DSO). Before using the pressure transducers in the experiments, all of the pressure transducers are tested with a standard gauge to make sure that the error of the pressure measurement is minimal. This also provides the linear relationship between output (voltage) and pressure (bar).

The tube of pressure transducer is connected into a bottom plate of the arc chamber. There are three locations to install the pressure transducers in the arc chamber: in the fixed contact region, behind the moving contact and behind the arc stack. To measure the pressure behind the arc stack, an extra thermal plastic sheet is used to cover the tube and pressure transducer which install behind the vent plate next to the arc stack.

The short circuit arc current is used as a trigger signal for the pressure transducers to capture the pressure signal from the arc chamber. The output signal is recorded in a memory of the digital storage oscilloscope (DSO) as a real time function from the arc ignition until the arc extinguish.

Afterward, the program "Flexgav4.base" downs load the data stored on the memory of the Digital Storage Oscilloscope (DSO) into the hard disk of computer (PC) via interface bus for further analysis.

3.11.5 Associated equipment operation

After the SPACEBAR is pressed the control computer energises the solenoid. The hammer then moves passes the optical sensor. This trigger sends a signal to the counter in the personal computer (PC) to commence a count down.

The period from the hammer triggers the sensor and impacting the contact mechanism to execute the contact opening delay (t_{cod}) as required. The computer (PC) calculates the time period (t) from that period minus the contact opening delay time t_{cod}. This period (t) is loaded into I/O card to count down the time.

When the counter reaches zero, the trigger starts to produce the fault current from capacitor discharge bank (CDB). Then the hammer impacts the contact mechanism. The contact is opened and the arc is drawn. The data of the light intensity acquisition is recorded in the Arc Imaging System (AIS).

Then, the digital storage oscilloscope (DSO) records the fault current and the fault voltage profiles. The computer (PC) turns off the high power thyristor in the capacitor discharge bank. The sequentially Flexible Test Apparatus (FTA) is isolated from the capacitor discharge bank (CDB).

Then the data stored on the memory of the Arc imaging System (AIS) and the Digital Storage Oscilloscope (DSO) is downloaded into the hard disk of computer (PC) via interface bus for further storage and analysis.

3.12 Method for evaluating arc root contact time

The optical data from the AIS allows a study of the arc root contact time on the fixed and moving contact individually. The arc position is calculated from the numerical data, the position of the Centre of Intensity (COI) of the arc for each sample period from the light intensity distribution of the whole arc.

The products of the light intensity and the position were summed over the whole array, and then divided by the sum of the total light intensity,

$$x = \frac{\sum I_i X_i}{\sum I_i}, \, y = \frac{\sum I_i Y_i}{\sum I_i}.$$

The result of the calculation gives the X and Y co-ordinate of the average light intensity for each time sample period. The position of the arc is subsequently defined as the position of the centre of intensity of the light at any point in time [2]. The position of the arc root can be determined by the position of the light intensity along the selected fibres along the arc runners as shown in Figure 4.2, Chapter 4.

The arc voltage and arc root trajectories are shown in Figure 3.21 to 3.22. The arc voltage is shown as the lower trace. The cathode root and anode root on the moving contact and on the fixed contact as shown in the upper trace are studied individually. To allow a full analysis of parameters, the arc root contact time on the fixed or moving contact is defined as follow:

3.12.1 Arc root contact time on the moving contact

The anode and cathode arc root contact time on the moving contact is defined as the time difference between the start of the arc (The point that the arc voltage wave form rises up rapidly) and at that point that the arc root displacement passes a 10 mm as shown in Figure 3.21.

The arc root contact time on the moving contact is shown to start moving away from the contact region at approximately 1550 µs. The contact opened and the arc occur at the time 468 µs. Therefore, the arc root contact time on the moving contact is approximately 1550-468=1082 µs.

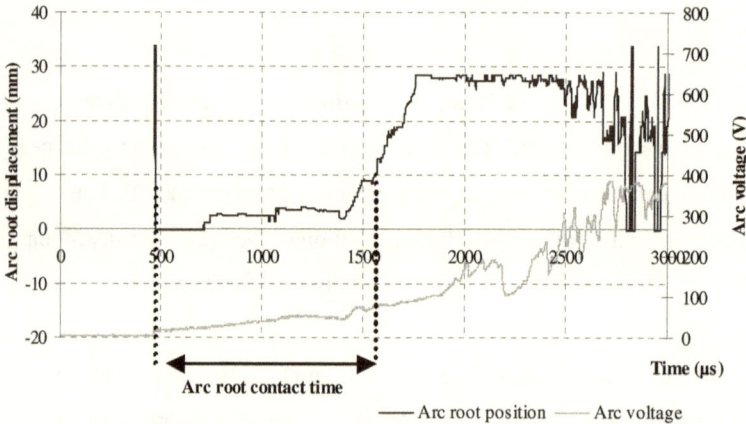

Figure 3.21: Arc root contact time on the moving contact

3.12.2 Arc root contact time on the fixed contact

The anode and cathode arc root contact time on the fixed contact is defined as the time between the start of the arc (The point that the arc voltage wave form rises up rapidly) and the time at that the arc root displacement moved away from the region at 0 mm as shown in Figure 3.22.

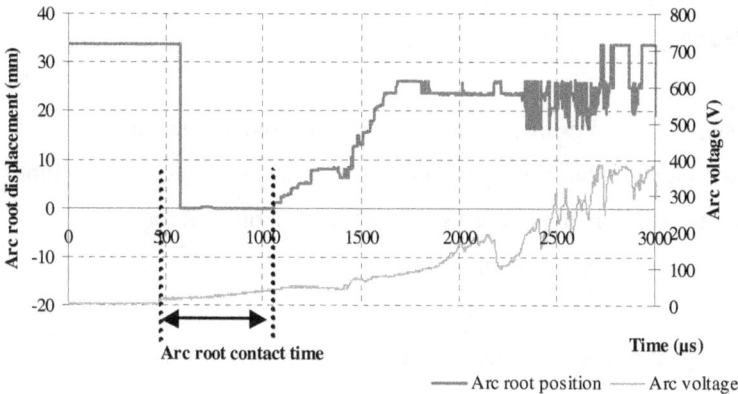

Figure 3.22: Arc root contact time on fixed contact

The arc root contact time on the fixed contact is shown to start moving away from the contact region at approximately 1100 μs. The arc is started at the time 468 μs. The arc root contact time on the fixed contact is approximately 1100-468=632 μs.

3.13 Summary

The Flexible Test Apparatus (FTA) is used to simulate the operation of a miniature circuit breaker when a short circuit current occurs. The Arc imaging System (AIS) is employed to record the optical data from the arc chamber in the Flexible Test Apparatus (FTA).

Two pressure transducers, the SX series of piezo resistive pressure sensor, are used to monitor the gas pressure in the arc chamber. The optic fibre spectrometer, a computer (PC) based S2000 plug in miniature optical fibre, is installed to measure the spectral data during the arcing.

The commercial package software programme OOOB is used to detect the signal of the arc spectrum. Flexgav4.bas is the programme to run the test sequences operation. Arcimage.bas is the program to present a series of the arc motion. Rootplot.bas is programme to analyse the experimental results.

The arc chamber geometry and contact opening velocity are modified. The velocity of the solenoid movement is controlled by the voltage across the armature of the stepping motor. To limit the gap behind the moving contact, a piece of Macro ceramic is used to close the gap behind the moving contact.

CHAPTER 4

EXPERIMENTAL RESULTS

4.1 Introduction

The experimental results in this chapter are divided into three main sections. The first section presents the experimental results of the arc root motion. The influences of the contact configuration, supply polarity, arc chamber venting, short circuit current level, contact opening velocity and the gap behind the moving contact are investigated.

The second section shows the experimental results of pressure in the arc chamber. The influences of the contact opening velocity, the gap behind the moving contact and the arc chamber materials are also considered.

The third section presents the experimental results of the arc spectrum. The influences of the contact opening velocity, contact materials and arc chamber materials are examined.

4.2 Arc root motion monitoring

A study of the arc root movement from the contact region concentrates on monitoring the arc root contact time at the point at which the arc root moves from the contact region.

The analogue light levels are recorded as digital signal by 6 bit A-D converters in the Arc Imaging System (AIS). The light intensity of the short circuit is digitised into 64 discrete levels of relative intensity. The relative light intensity records as an array of the X and Y co-ordinates over the position of optical fibres in the arc chamber.

The shape of the arc can be presented using five dynamic threshold levels to compute contour levels of the light intensity distribution. The relative light intensity thresholds are divided into five divisions between the maximum light intensity and zero.

The arc contour images are generated in colour from low light level. It starts from the yellow colour, light green, light red, red and black. The black colour is the maximum light intensity. This gives advantages that the arc contour images can be generated at low light levels during the early state of the arc event.

Therefore, the same colour region could correspond to a different light intensity at a different duration of the arc. However, the arc images are useful to gain information on the arc motion in the early stages.

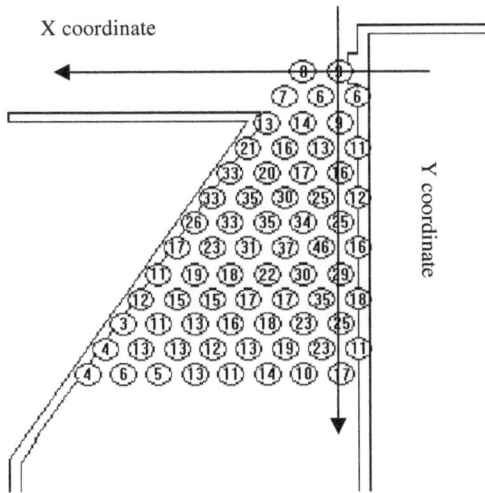

(a)

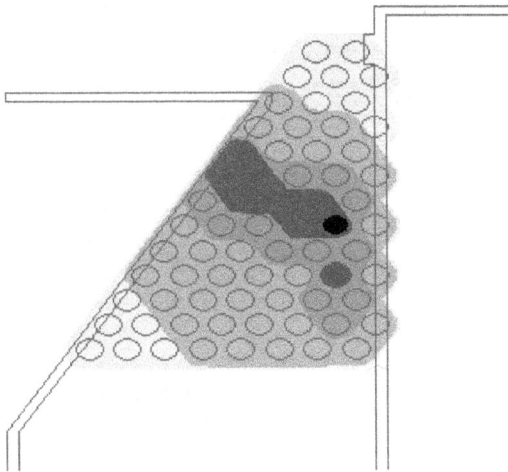

(b)

Figure 4.1: (a) The light levels recorded over the fibre optic array for one AIS sample period. Run 2030, t = 1360 μs. (b) Contour of arc imaging [2]

4.3 Arc root contact time

The arc root contact time is determined from the selected optical data as shown in Figure 4.2. The method for evaluating the arc root contact time is defined in section 3.12, Chapter 3.

There are many factors that affect the movement of the arc root from the contact region as presented in the literature review in Chapter 2. The studies of the arc root motion investigate cathode and anode arc root individually on the fixed and moving contact. The results presents with +/- 1 standard deviation. The first parameter considered is the Ag/C contact configuration.

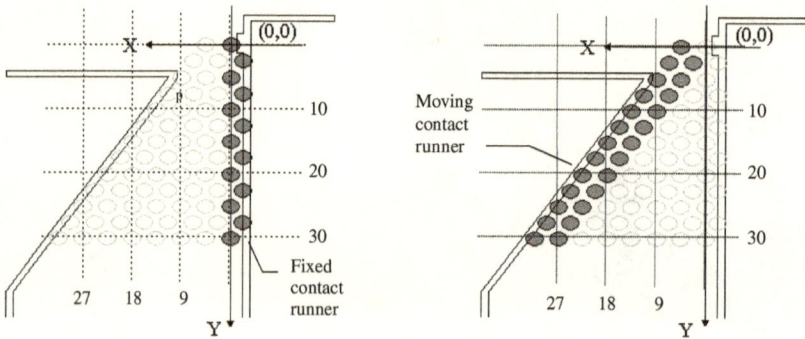

(A) Arc root on the fixed contact   (B) Arc root on the moving contact

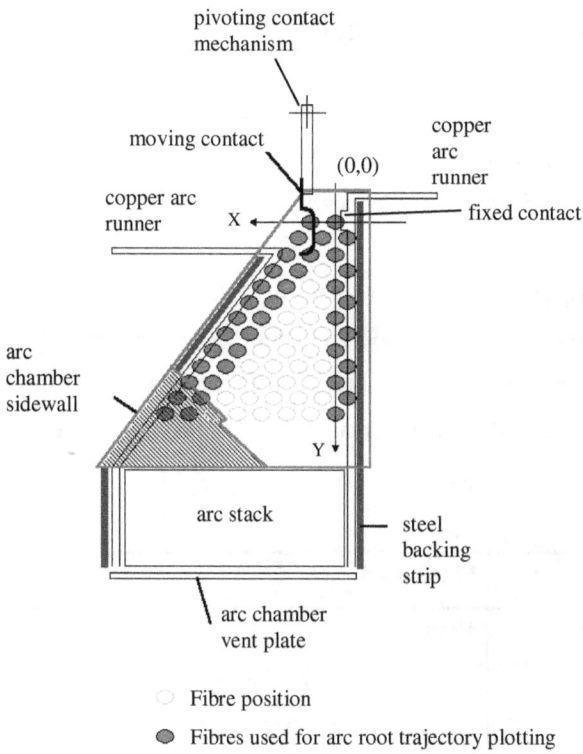

Figure 4.2: Arc chamber geometry. The circles identify optical fibre positions; the dark circles are positions used to monitor arc root motion [2,30]

4.3.1 Ag/C contact configuration

The influences of contact configuration on the arc root motion from the contact region have been investigated. A piece of Ag/C contact plate is welded on the moving and fixed contact, see Figure 3.19, Section 3.8, Chapter 3.

The arc voltage and arc root displacement are recorded to enable to specify the point at which the arc root starts to move off from the contact region. These experiments use plain Cu for the fixed contact. The pivot contact mechanism is used to operate the short circuit current. The summary of the experimental condition is shown in table 4.1.

Experimental Factors	Fixed level
Contact velocity (m/s)	10 m/s
Supply polarity	- Anode on the moving contact - Cathode on the fixed contact
Short circuit current (A)	2000 A
Contact material	Ag/C step on the moving/fixed contact
Arc chamber venting	Opened, choked, closed
Arc chamber material	Ceramic

Table 4.1: Experimental conditions for Ag/C on the moving contact

The experimental results of the arc root displacement and arc voltage are shown in Figure 4.3. Figure 4.4 presents the experimental results of the arc root contact time when Ag/C is on the fixed and moving contact.

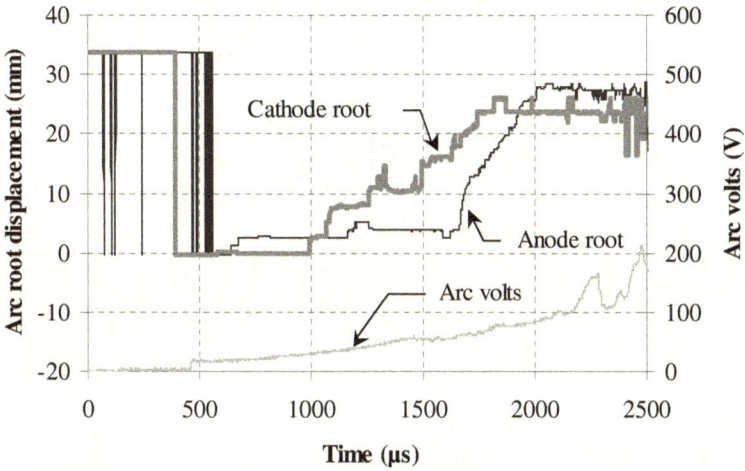

Arc chamber venting: opened

Arc chamber venting: choked

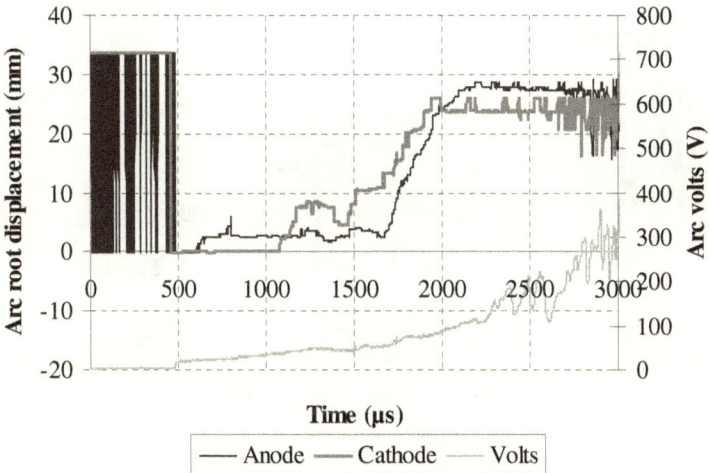

Arc chamber venting: closed

(a) Ag/C on the moving contact

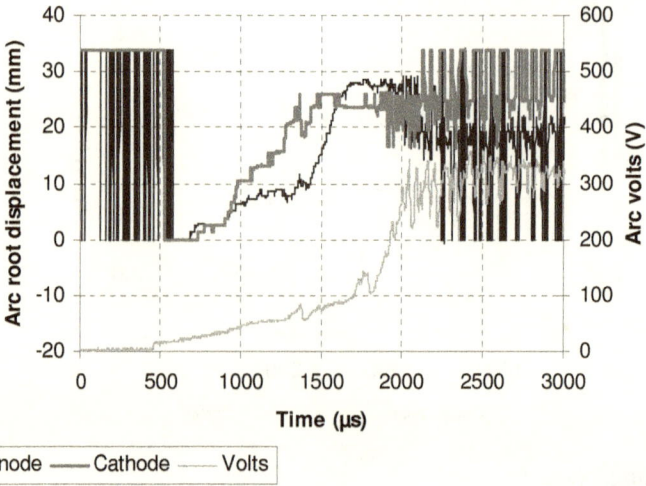

Arc chamber venting: opened

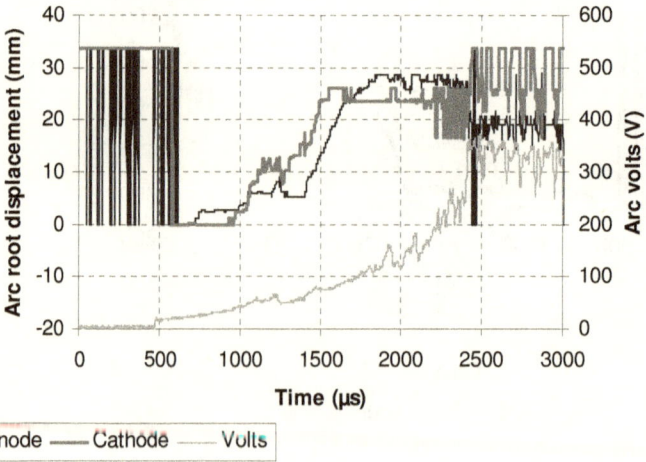

Arc chamber venting: choked

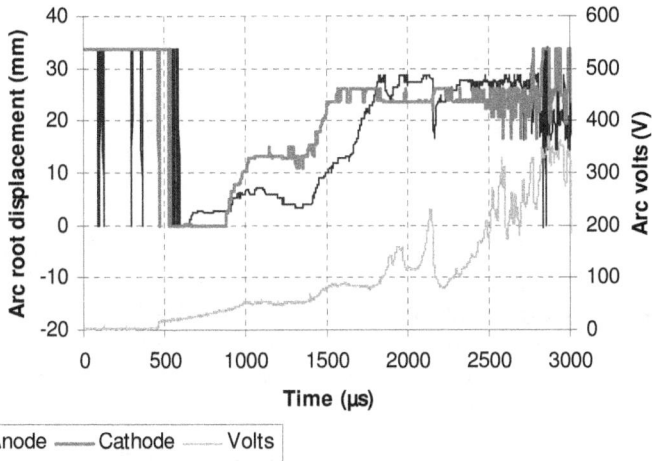

Arc chamber venting: closed

(b) Ag/C on the fixed contact

Figure 4.3: Arc root displacement when Ag/C on the moving (a) and fixed contact (b) with opened, choked and closed arc chamber venting, contact opening velocity of 10 m/s, short circuit current 2000 A, ceramic arc chamber material

The waveforms of the arc root displacement and arc voltage when Ag/C is on the moving and fixed contact is shown in Figure 4.3. In all case the anode root (on the moving contact) is delayed longer than the cathode root (on the fixed contact).

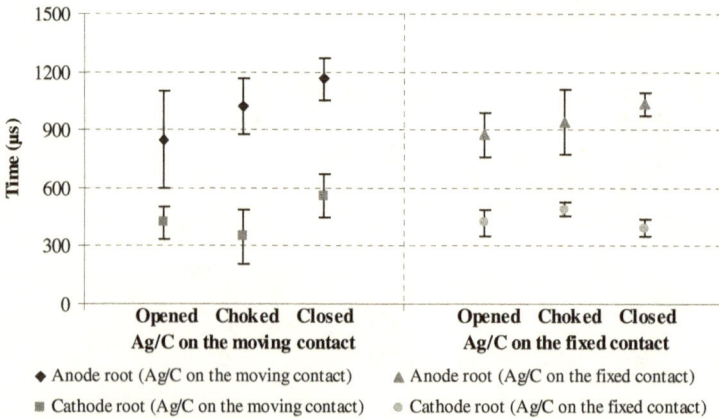

Figure 4.4: Arc root contact time when Ag/C on the moving and fixed contact with arc chamber venting: opened, choked and closed, contact opening velocity of 10 m/s, short circuit current 2000 A, ceramic arc chamber material

The anode and cathode root contact times when a piece of Ag/C contact material is welded on the moving and fixed contacts are shown in Figure 4.4. The experimental results, when Ag/C is on the moving contact shows that the anode root on the moving contact delays longer on the moving contact region than the cathode root on the fixed contact.

This may be caused from the piece of Ag/C on the tip of the moving contact. On the other hand, when Ag/C is on the fixed contact the anode root contact time is longer than the cathode root contact time. The anode root contact time when Ag/C on the fixed is shorter than Ag/C on the moving contact.

With Ag/C on the moving contact, a piece of Ag/C removed off from the moving contact after a very short time. This leads to the conclusion that the geometry of contact with a piece of Ag/C on the moving contact is not suitable for commercial use in miniature circuit breakers.

4.3.2 Polarity effect

The effects of the arc chamber venting are clearly shown in Figure 4.4. The area venting has an significant influence on the anode root contact time on the moving contact. The arc root contact time decreases as the area of the arc chamber venting is decreased on both Ag/C on the moving and fixed contact. The arc chamber venting shows an insignificant effect on the cathode root on the fixed contact for both Ag/C on the moving and fixed contact.

However, it is not clear whether the arc chamber venting has an influence solely on the anode on the moving contact or on the cathode on the moving contact as well. Moreover, there have been numerous studies of the cathode root motion since the cathode root is generally thought to dominate the motion of the arc root from the contact region.

To investigate this, experiments with Ag/C is on the fixed contact and reversed supply polarity are undertaken. Contact opening velocity is 10 m/s and ceramic is used for the arc chamber material.

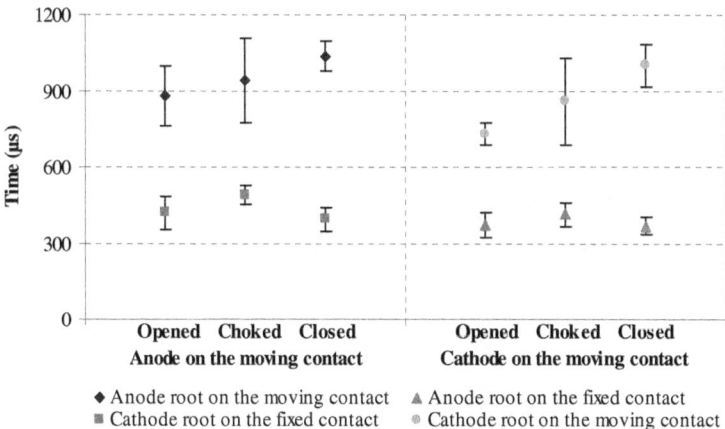

◆ Anode root on the moving contact	▲ Anode root on the fixed contact		
■ Cathode root on the fixed contact	● Cathode root on the moving contact		

Figure 4.5: Effect of polarity on arc root contact time when anode and cathode is on the moving contact, Ag/C on the fixed contact, contact opening velocity 10 m/s, ceramic arc chamber material, arc chamber venting: opened, choked and closed

The experimental results of the anode and cathode root contact time when the anode is on the fixed contact and the cathode is on the moving contact are shown in Figure 4.5. The experimental results indicate that the anode root contact time on the moving contact is longer than the cathode root contact time on the moving contact.

Both anode and cathode on the fixed are not significantly affected by the arc chamber venting area. The anode and cathode root contact time on the moving contact with opened arc chamber venting is lower than the choked and closed condition.

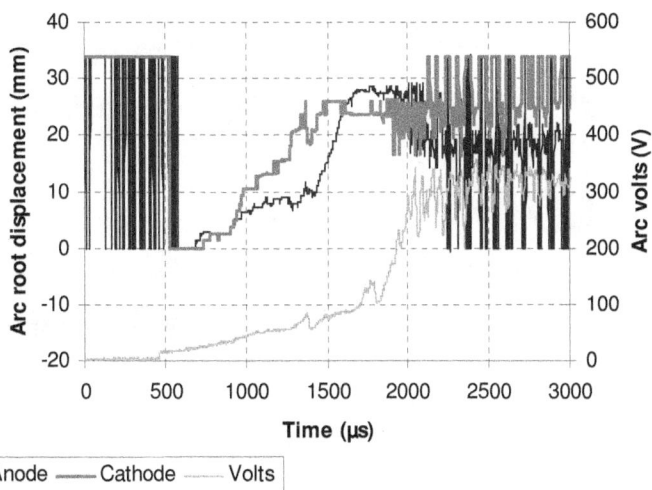

Arc chamber venting: opened

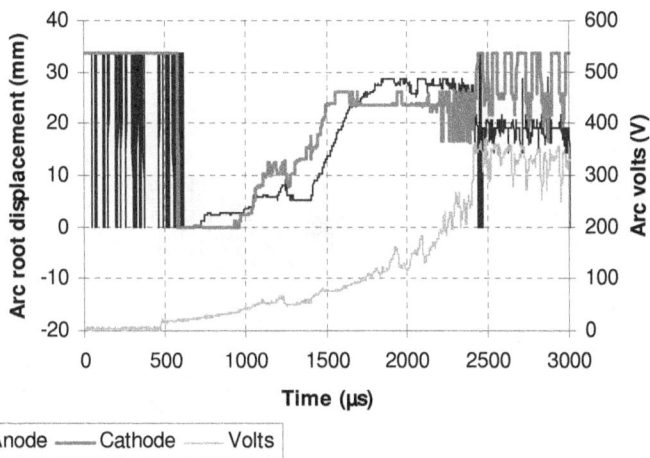

Arc chamber venting: choked

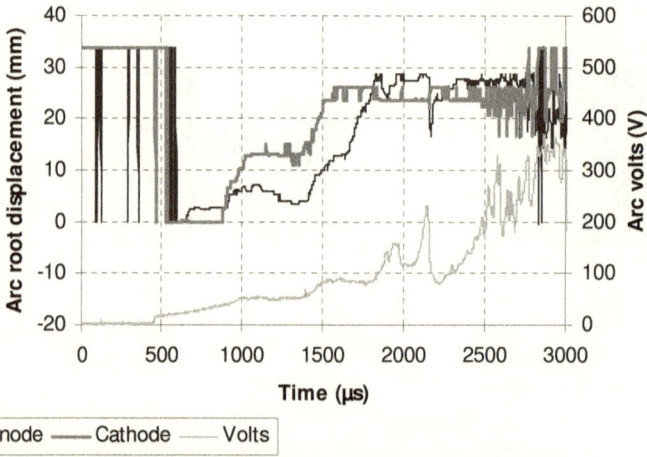

Arc chamber venting: closed

(a) Anode is on the moving contact , cathode is on the fixed contact

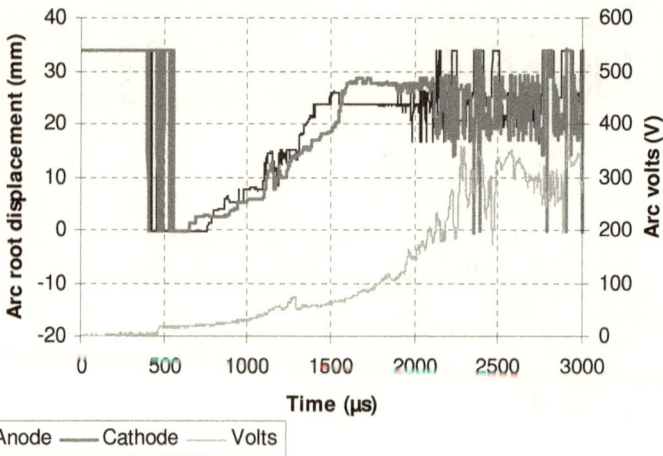

Arc chamber venting: opened

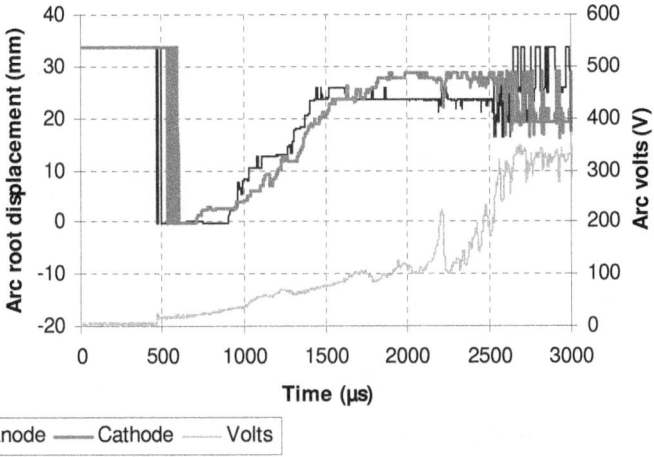

Arc chamber venting: choked

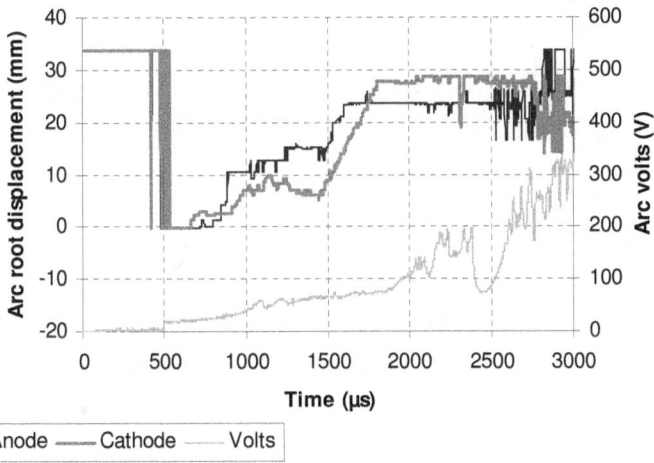

Arc chamber venting: closed

(b) Cathode is on the moving contact, anode is on the fixed contact

Figure 4.6: Arc root displacement when the anode (a) and cathode (b) on the moving contact and opened, choked, closed arc chamber venting, Ag/C on the fixed contact, contact opening velocity 10 m/s and ceramic arc chamber material

Figure 4.6(a) shows the waveform of the arc root displacement from contact region to the arc stack with opened, choked and closed arc chamber venting when anode is on the moving contact.

When the anode is on the moving contact and cathode is on the fixed contact, the arc root delays in the contact region longer with the closed arc chamber venting than the opened and choked arc chamber venting.

The anode and cathode root displacement shows the arc re-strike in the arc chamber with the closed arc chamber venting. The anode root contact time on the moving contact is longer than the cathode root contact time on the fixed contact.

The waveforms of the arc root displacement when anode is on the fixed contact and cathode is on the moving contact are shown in Figure 4.6 (b). With opened arc chamber venting, both anode and cathode root moves from the contact region earlier than choked and closed.

4.3.3 Short circuit current level

During a short circuit fault, an electric arc is drawn between opening contacts. The current through the conductors of the MCBs generates a magnetic field in the arc chamber.

This magnetic force acts to force the arc away from the contact region along arc runners into an arc stack and pass through the arc chamber venting behind the arc stack. The magnetic force is a function of the short circuit arc current. Therefore, the influences of the short circuit arc current on the arc root contact time are considered.

These experiments have been studied with both anode and cathode on the moving contact. The peak short circuit current level investigations here are 500 A, 1400 A and 2000 A.

The choked condition is selected because the area venting of this condition is similar to the commercial MCB's. The Ag/C step is fitted on the fixed contact and the arc chamber material is ceramic.

The experimental results of the arc root contact time with short circuit current levels at 500 A, 1400 A and 2000 A are shown in Figure 4.7.

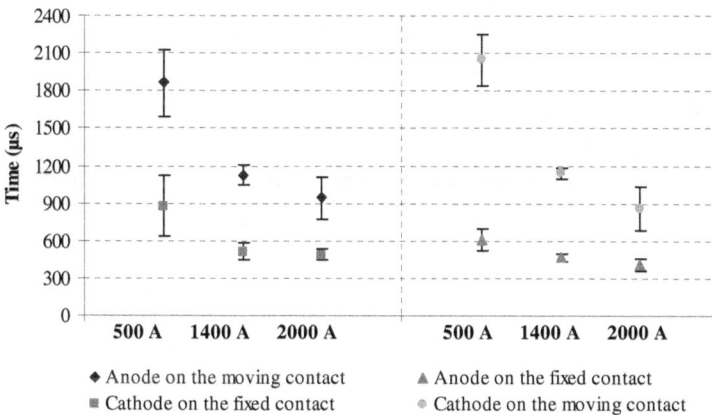

Figure 4.7: Arc root contact time and short circuit current level with contact opening velocity 10 m/s, Ag/C on the fixed contact, ceramic arc chamber and choked arc chamber venting

Figure 4.7 shows the experimental results of the anode and cathode root contact time with the short circuit current levels 500A, 1400A and 2000A. The results show that the arc root contact time decreases as the short circuit current is increased.

The anode root delays on the moving contact region longer than the cathode root on the fixed contact. The results are similar to the cathode root on the moving contact and the anode root on the fixed contact.

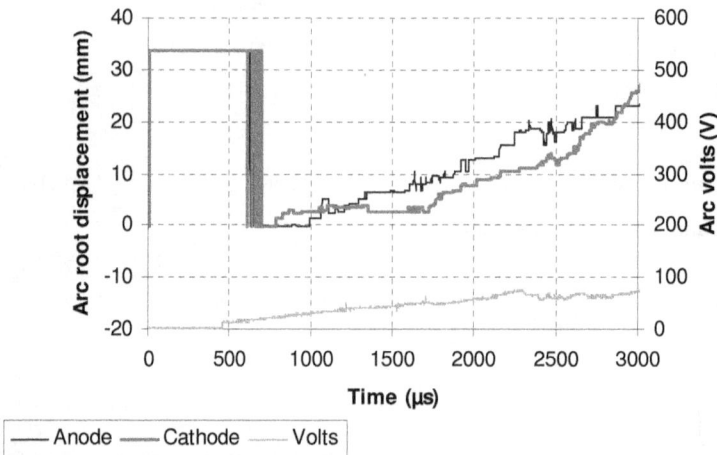

(a) Short circuit current 500 A

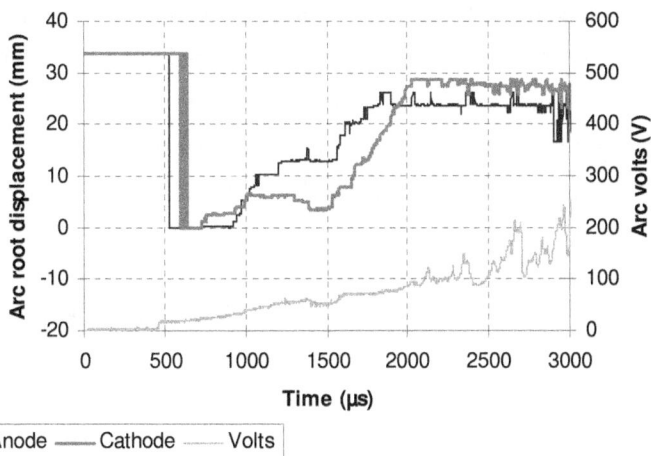

(b) Short circuit current 1400 A

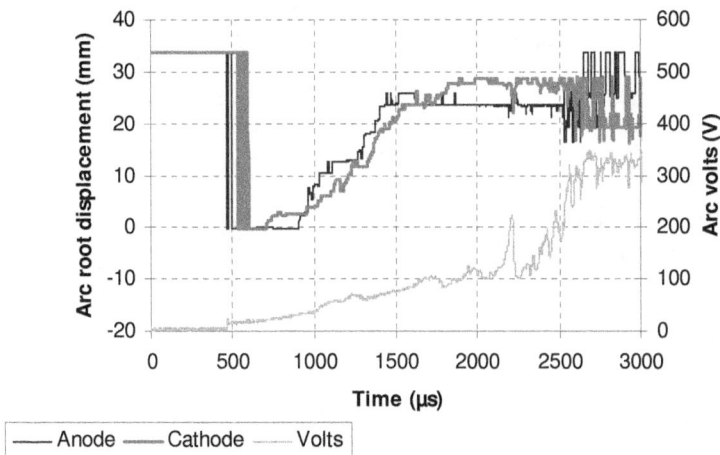

(c) Short circuit current 2000 A

Figure 4.8: Arc root displacement when short circuit current 500A, 1400A and 2000A, cathode on the fixed contact and anode on the moving contact, contact opening velocity 10 m/s, ceramic arc chamber and choked arc chamber venting

Figure 4.8 shows the waveform of the arc root displacement of the short circuit current level 500A, 1400A and 2000A. For short circuit current level 500A, both anode and cathode root contact delay longer in the contact region before it moves slowly off from the contact region towards into the arc chamber. At the short circuit current level 2000A, the arc root moves from the contact region faster than that of 1400A and 500A.

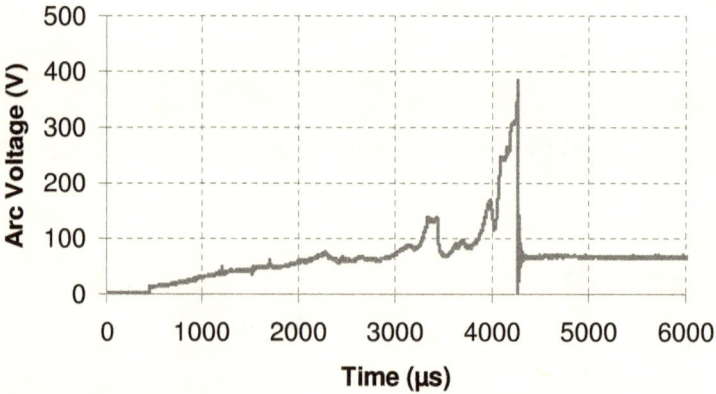

Short circuit current 500 A

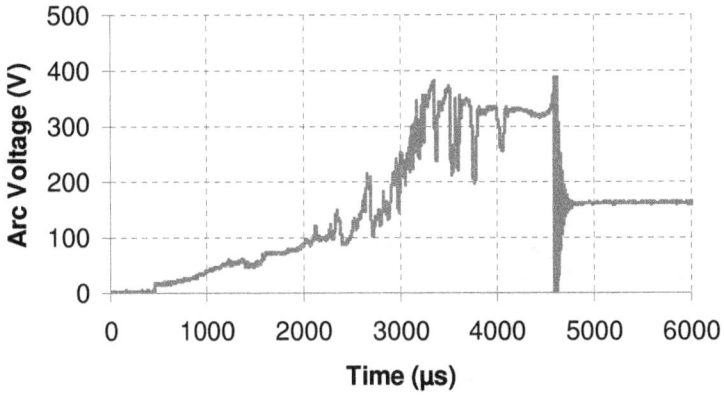

Short circuit current 1400 A

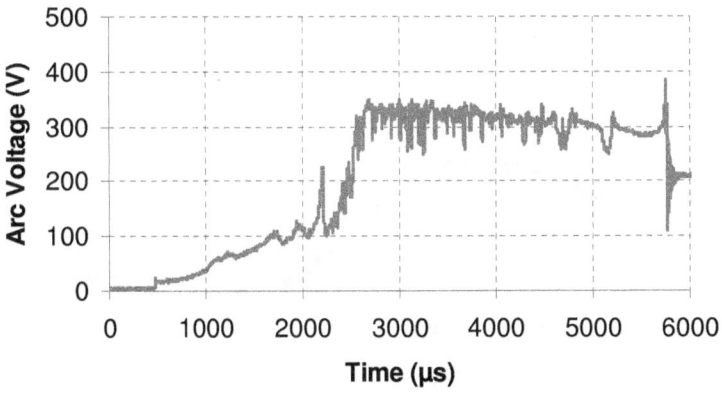

Short circuit current 2000 A

(a) Arc voltage

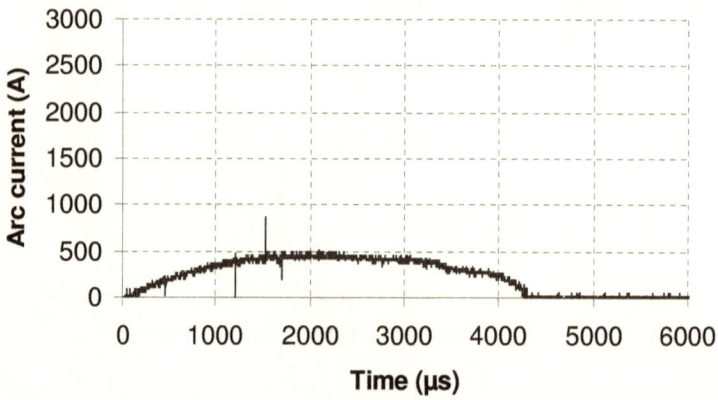

Short circuit current 500 A

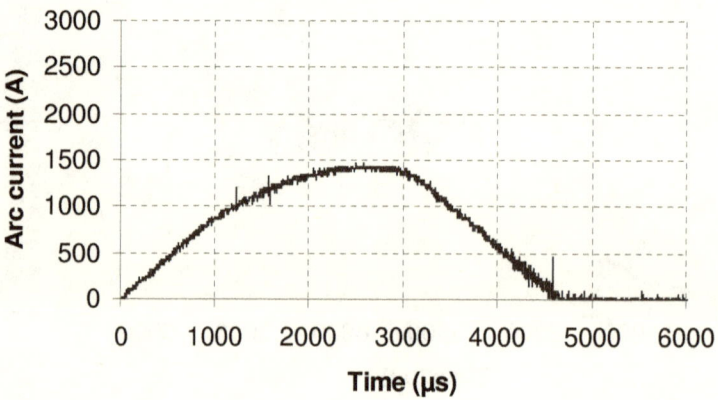

Short circuit current 1400 A

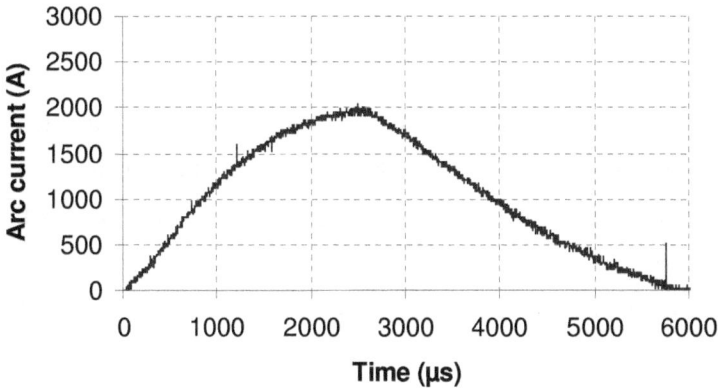

Short circuit current 2000 A

(b) Arc current

Figure 4.9: (a) Arc voltage and (b) Arc current with short circuit current level 500A, 1400 A and 2000 A, cathode on the fixed contact and anode on the moving contact, contact opening velocity 10 m/s, ceramic arc chamber and choked arc chamber venting

The waveforms of the arc voltage with short circuit current level 500A, 1400A and 2000A are shown in Figure 4.9 (a). The arc voltage for short circuit current level 500A rises up slightly until the arc root reaches the arc stack at the period of 4200µs.

For the short circuit current level 1400A, the arc root reaches the arc stack earlier than the short circuit current level 500A. The waveform of the arc voltage for the short circuit current level 2000A show the arc re-strike before the arc root moves towards the arc stack.

The waveforms of the arc current with the short circuit current level 500A, 1400A and 2000A are shown in Figure 4.9 (b). The total period of the arc for the short circuit current level 500A and 1400A is about 4500 µs but for the short circuit current level

2000A, the total period of the arc is up to 5500 μs. The maximum short circuit current is about 2500 μs.

The short circuit current shows a significant influence on the movement of the arc root from the moving contact region. Both anode and cathode root contact times on the moving and fixed contact decrease as the short circuit peak current is increased.

4.3.4 Contact opening velocity

It has been shown that the influence of the arc chamber venting and short circuit current level on the arc root motion has more significant effects on the moving contact than the fixed contact, as present in section 4.3.2 and section 4.3.3.

From the literature review in section 2.4.4, when the contact opening velocity is above 6 m/s the arc root contact time is reduced to a minimum value. However, it is not clear whether the influence of contact opening velocity effects both anode and cathode root.

The mechanism to actuate the moving contact in these experiments is the pivot system. The contact opening velocity can be adjusted to observe the events of the arc from 1 m/s up to 10 m/s. The effects of the contact opening velocities have been examined with various speed of contact from 1 m/s, 4 m/s, 5.5 m/s and 10 m/s.

The arc chamber venting is in the choked condition. A ceramic is used as the arc chamber material. The short circuit current level is set at 2000 A. The power supply anode is on the moving contact and the cathode is on the fixed contact

The experimental results of the arc root contact time with contact opening velocities of 1 m/s, 4 m/s, 5.5 m/s and 10 m/s are shown in Figure 4.10.

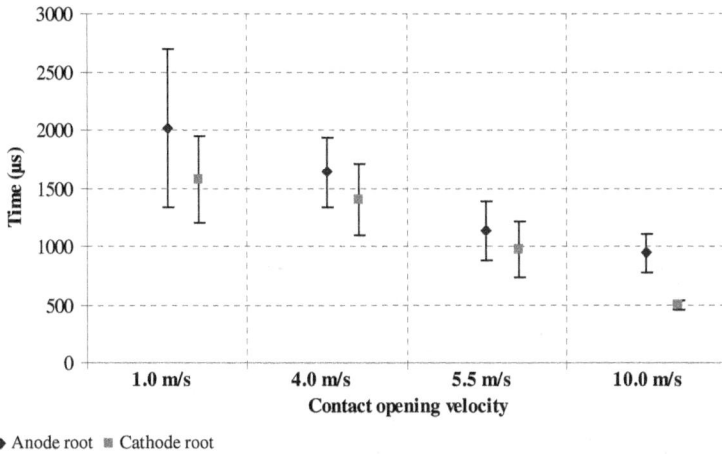

Figure 4.10: Arc root contact time with contact opening velocity 1 m/s, 4 m/s, 5.5 m/s and 10 m/s, cathode on the fixed contact and anode on the moving contact, short circuit current 2000 A, ceramic arc chamber and choked arc chamber venting

Figure 4.10 presents the experimental results of the anode and cathode root contact time under the study of the effects of the contact opening velocity. The experimental results show that both anode and cathode root contact time decreases as the velocity of opening contact is increased. The arc root contact time when the contact opening velocity of 1 m/s is higher than that of 4.0 m/s, 5.5m/s and 10 m/s.

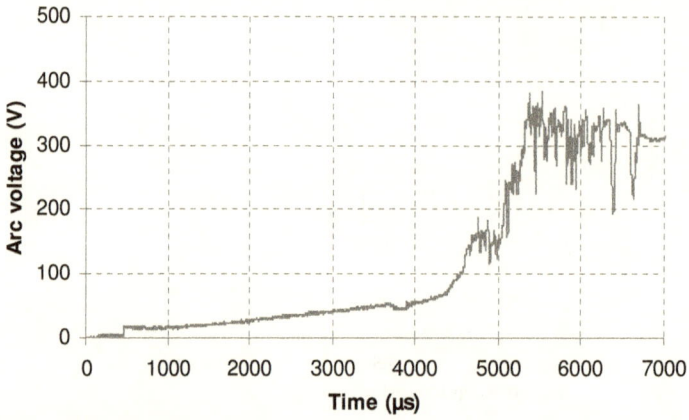

Contact opening velocity 1 m/s

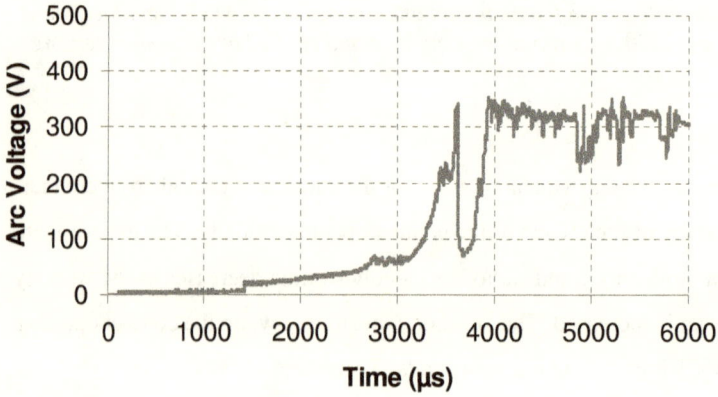

Contact opening velocity 4 m/s

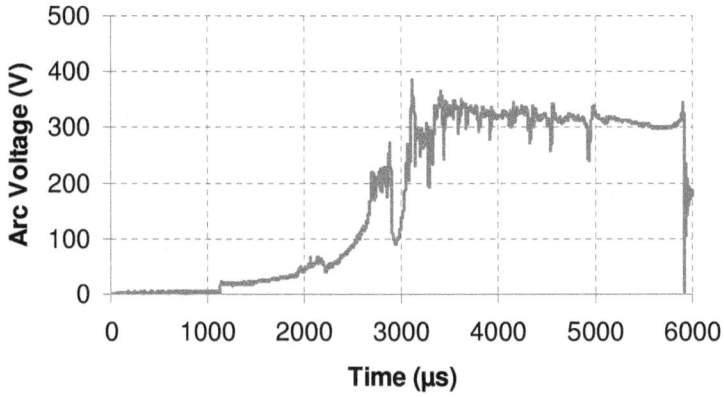

Contact opening velocity 5.5 m/s

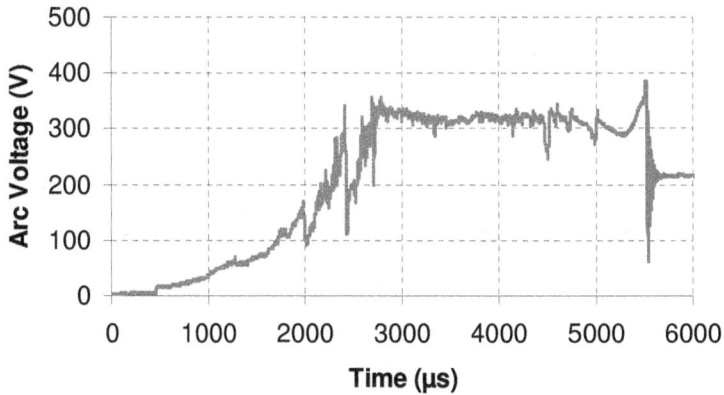

Contact opening velocity 10 m/s

(a) Arc voltage

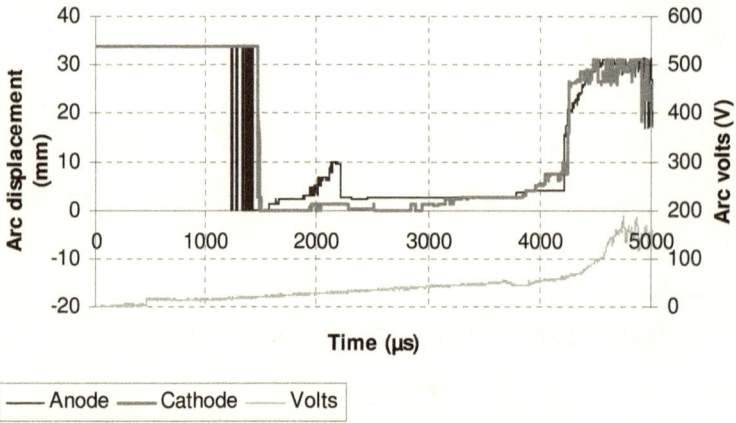

Contact opening velocity 1 m/s

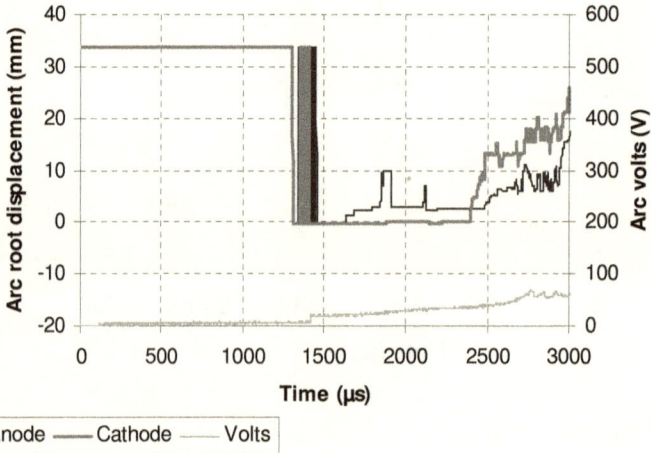

Contact opening velocity 4 m/s

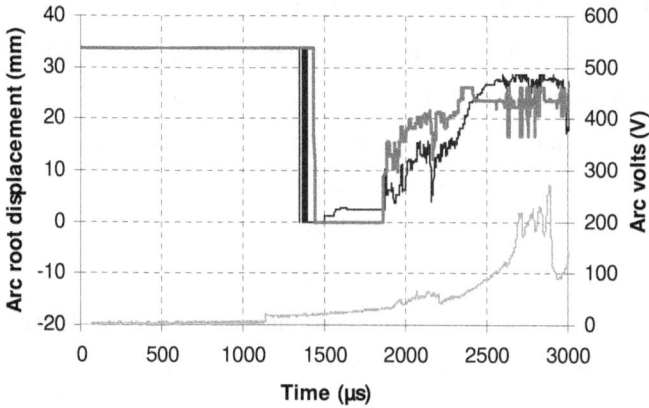

Contact opening velocity 5.5 m/s

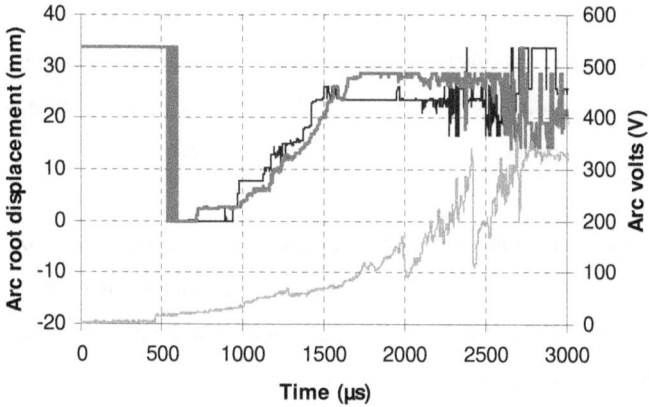

Contact opening velocity 10 m/s

(b) Arc root displacement

Figure 4.11: (a) Arc voltage and (b) Arc root displacement when contact opening velocity 1 m/s, 4 m/s, 5.5 m/s and 10 m/s, cathode on the fixed contact and anode on the moving contact, short circuit current 2000 A, ceramic arc chamber and choked arc chamber venting

The waveforms of the arc voltage with contact opening velocity of 1 m/s, 4 m/s, 5.5 m/s and 10 m/s are shown in Figure 4.11 (a). At contact opening velocity 1 m/s, the profile of the arc voltage increases slightly before the arc root arrives at the arc stack at 4500 μs. The period of the arc root moves from the contact region into the arc stack for contact opening velocity of 1 m/s is longer than that of 4 m/s, 5.5 m/s and 10 m/s.

The arc re-strike is clearly shown in front of the arc stack in the arc chamber for contact opening velocity of 1 m/s, 4 m/s and 5.5 m/s. The total period of the arc is about 6000 μs for contact opening velocity 5.5 m/s and 10 m/s. The arc period is more than 6000 μs for contact opening velocity 1 m/s and 4 m/s.

The waveforms of the arc root displacement with contact opening velocity 1 m/s, 4 m/s, 5.5 m/s and 10 m/s are shown in Figure 4.11 (b). The arc root delays to commute off from the contact longer at contact opening velocity 1 m/s than that of 4 m/s, 5.5 m/s and 10 m/s. At contact opening velocity more than 5.5 m/s, the arc root moves directly towards into the arc stack after leaving the contact region.

The experimental results show that the arc root on the moving and the fixed contact decreases as the contact opening velocity is increased. It is indicated that the contact opening velocity has a significant influence on the mobility of the arc root. The anode root moves off from the moving contact slower than the cathode root commutes from the fixed contact.

4.3.5 Gap behind the moving contact

The experimental results from Section 4.3.2, the arc chamber venting has a strong effect on both anode and cathode root on the moving contact. Considering the arc chamber geometry, the arc chamber venting behind the arc stack is not only the vent in the arc chamber, a gap behind the moving contact also can consider as a vent.

To investigate the influence of the gap behind the moving contact, the geometry of the arc chamber is modified. A piece of macro ceramic is added to limit the venting area behind the moving contact. The arc chamber venting is settled as opened, choked and closed.

Two experiments, opened and closed the gap behind the moving contact, are design to investigate the influence of the gap behind the moving contact on the arc root contact time. The detail of the specification of the experimental condition is shown in table 4.2.

The experimental results of the arc root contact time when the gap behind the moving contact is closed and opened are shown in Figure 4.12.

Experimental Factors	Fixed level
Contact velocity (m/s)	10 m/s
Supply polarity	- Anode on the moving contact - Cathode on the fixed contact
Short circuit current (A)	2000 A
Contact material	Ag/C step on the fixed contact
Arc chamber venting	Opened, choked, closed
Arc chamber material	Ceramic
Gap behind the moving contact	Opened/closed

Table 4.2: Experimental conditions to study the gap behind the moving contact

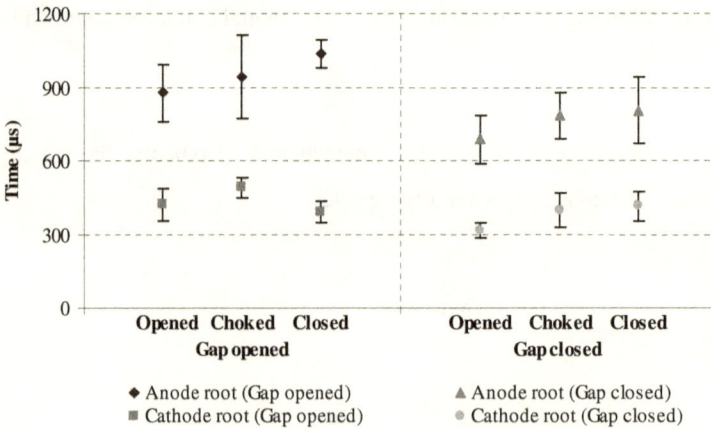

Figure 4.12: Arc root contact time when opened and closed the gap behind the moving contact, contact opening velocity 10 m/s, anode on moving contact and cathode on fixed contact, short circuit current 2000 A, ceramic arc chamber, arc chamber venting: opened, choked and closed

The experimental results of the cathode and anode root contact time with closed and opened the gap behind the moving contact is shown in Figure 4.12. When opened the gap behind the moving contact, the anode root on the moving contact delays longer than the cathode root on the fixed contact.

The anode root contact time on the moving contact has a significant effect on the opened gap behind the moving contact. On the other hand, the cathode on the fixed contact shows no significant influence from the gap.

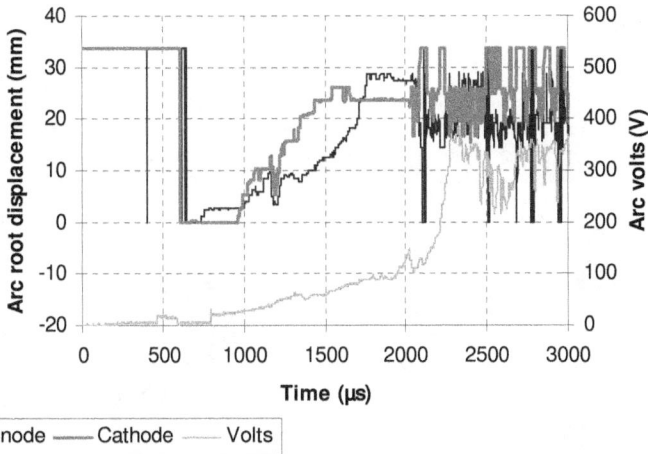

Arc chamber venting: opened

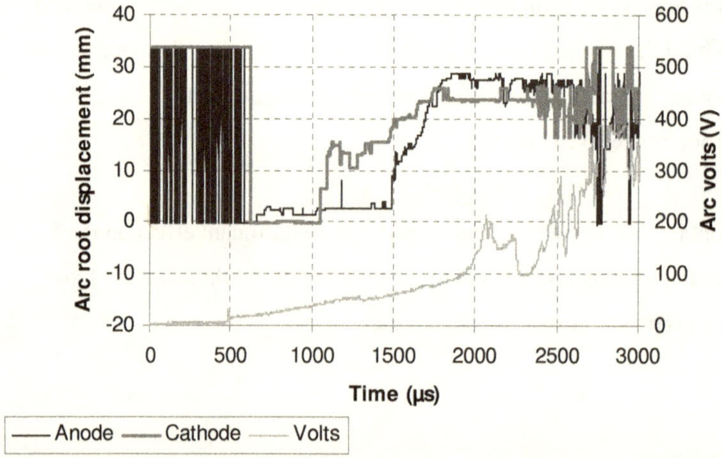

Arc chamber venting: choked

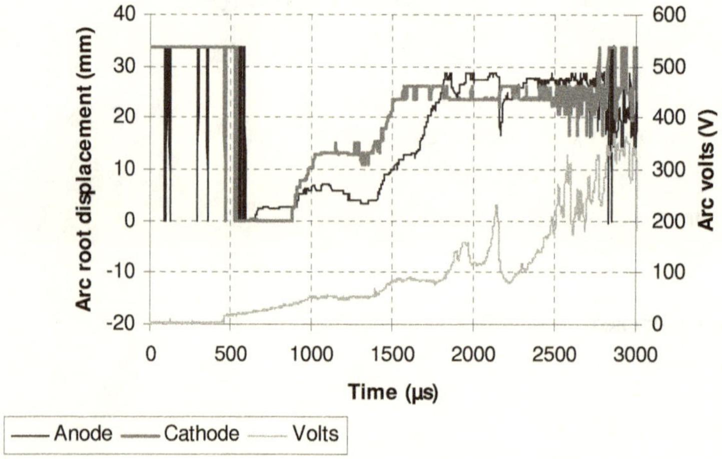

Arc chamber venting: closed

(a) Gap opened

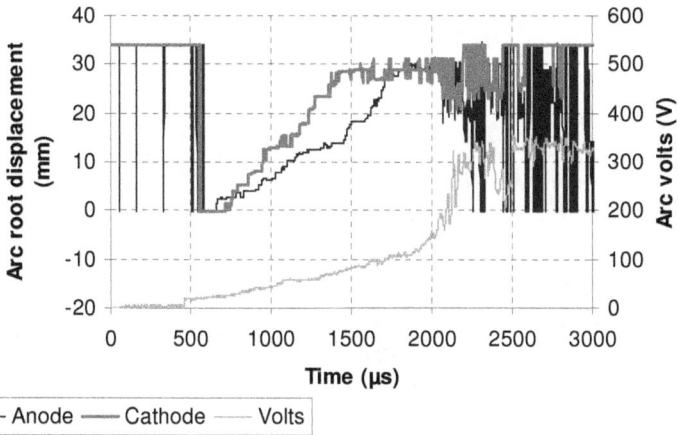

Arc chamber venting: opened

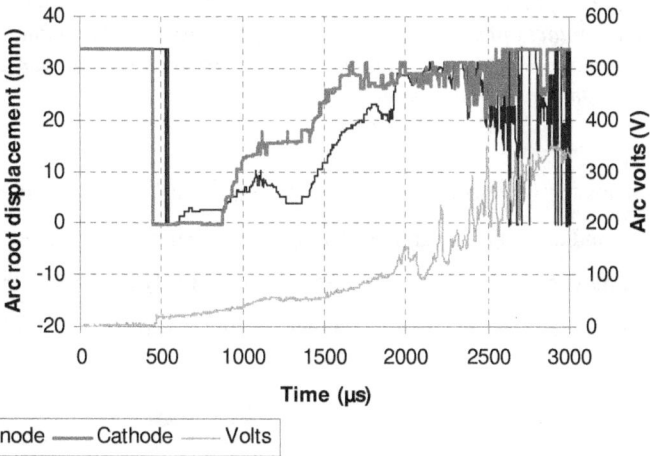

Arc chamber venting: choked

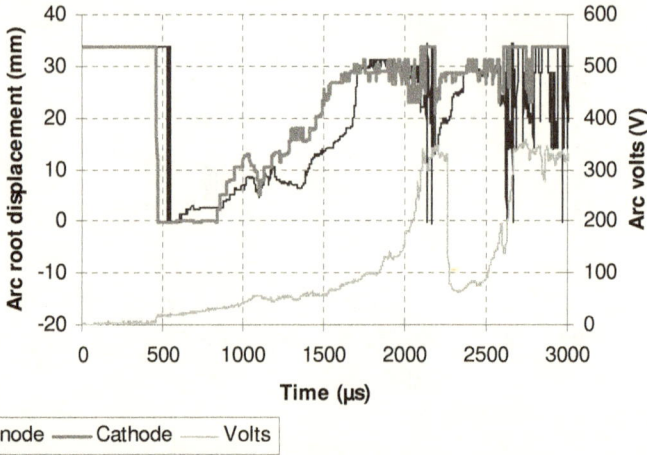

Arc chamber venting: closed

(b) Gap closed

Figure 4.13: Arc root displacement when opened (a) and (b) closed the gap behind the moving contact, contact opening velocity 10 m/s, anode on moving contact and cathode on fixed contact, short circuit current 2000 A, ceramic arc chamber, arc chamber venting: opened, choked and closed

Figure 4.13 (a) presents the waveform of the arc voltage and arc root displacement with opened, choked and closed arc chamber venting. With opened arc chamber venting, the arc root moves towards into the arc stack smoothly. The period of the arc root stays in the contact area for opened arc chamber venting is shorter than choked and closed.

The anode root contact time on the moving contact is higher than the cathode root on the fixed contact. The arc root displacements with opened, choked and close arc chamber venting and closed the gap behind the moving contact are shown in Figure 4.13 (b).

With opened arc chamber venting, both of the arc root moves directly into the arc stack after leaving the contact region. Both anode and cathode root delay long to move from the contact region for the choked and closed condition.

It is shown that the gap behind the moving contact has an influence on the anode contact time. The arc root contact time with the gap opened is about 17-24% longer than with the gap is closed. On the other hand, the cathode root on the fixed contact shows no significant influence from the gap.

4.3.6 Summary of the arc root contact time

The arc root on the moving contact delays in the contact region longer than the fixed contact for both Ag/C on the moving and fixed contact. It is considered that the geometry of contact with a piece of Ag/C on the moving contact is not suitable for commercial use in miniature circuit breakers.

The influence of the arc chamber venting has an effect on the arc root on the moving contact. The arc root contact times increase as the vent area is decreased. The arc root contact time on the fixed contact is not significantly affected by the arc chamber venting.

The arc root contact time with the gap behind the moving contact opened is longer than the gap closed. The gap has a significant effect on the arc root contact time on the moving contact. Conversely, the arc root on the fixed contact shows no significant effect on this gap. The chamber and contact vents are related.

The short circuit current shows a significant effect on the arc root motion from the moving contact region. The arc root contact time decrease as the short circuit peak current is increased.

The contact opening velocity has a significant influence on the mobility of the arc root moving off from the contact region. The arc root contact time decreases as the contact opening velocity is increased. To understand these events the following sections consider the pressure in the arc chamber and arc spectrum.

4.4 Pressure in the arc chamber

From the experimental results in Section 4.3, the arc chamber venting, short circuit current level, contact opening velocity and the gap behind the moving contact affect the arc root motion. The influences of the gap behind the moving contact, arc chamber material and contact opening velocity on the pressure in the arc chamber are considered in this section.

4.4.1 The gap behind the moving contact

From the literature review in section 2.5.2, the pressure depends on the outlet cross section of the arc chamber. The pressure imposed in the arc chamber outlet can limit the radius of the arc.

The arc chamber venting behind the arc stack is the main vent in the arc chamber but the gap behind the moving contact is also considered as a vent. Although the dimension of the gap behind the moving contact is not as big as the area of the arc chamber vent, it has an important influence on the arc root contact time as described in Section 4.3.5.

Thus, the influences of the gap behind the moving contact on the pressure in the arc chamber have been investigated.

To investigate the effect of the gap behind the moving contact on the pressure in the arc chamber, two pressure transducers are used to measure the pressure at the point behind the arc stack and in the gap behind the moving contact. This experiment used choked condition as the arc chamber venting. The summary of experimental condition is shown in table 4.3.

Experimental Factors	Fixed level
Contact velocity (m/s)	10 m/s
Supply polarity	- Anode on the moving contact - Cathode on the fixed contact
Short circuit current (A)	2000 A
Contact material	Cu punch on the fixed contact
Arc chamber venting	Choked
Arc chamber material	Ceramic
Gap behind the moving contact	Opened

Table 4.3: Experimental conditions to study the gap behind the moving contact

The pressure profiles behind the arc stack and behind the moving contact are shown in Figure 4.14 and the arc root displacement is shown in Figure 4.15.

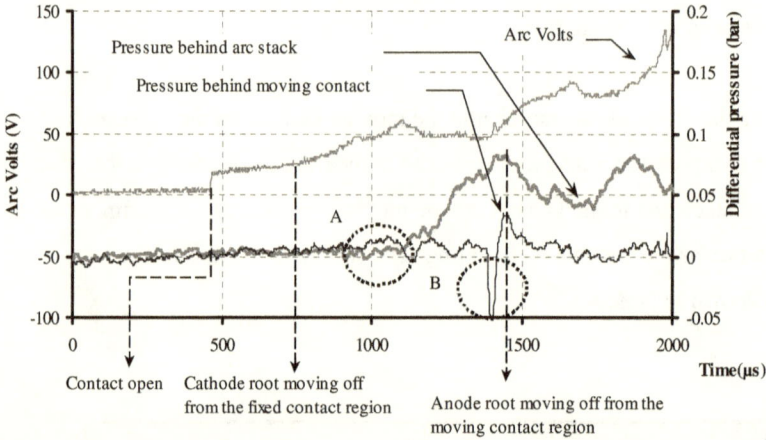

Figure 4.14: Differential pressure behind the arc stack and behind the moving contact of Cu punch contact material, 10 m/s, Anode on moving contact, 2000 A. peak current, choked, ceramic arc chamber, open gap behind the moving contact

Figure 4.14 shows the waveforms of pressure behind the moving contact and behind the arc stack. The results show that the pressure behind the moving contact starts to rise up earlier than the pressure behind the arc stack. However, the pressure in the gap behind the moving contact does not rise up until stability.

At the point at which the contact starts to open, both profiles of the pressure are constant. The time at which the anode root commutes from the moving contact is coincident with the maximum point of the pressure behind the arc stack.

The pressure behind the moving contact drops immediately approximately 0.05 bar, while the maximum point of the pressure behind the arc stack rises up to 0.08 bar.

At "A" in Figure 4.14, the pressure behind the contact starts to rise before the pressure behind the arc stack. When the arc roots move off from the contact region, the pressure behind the arc stack is higher than the pressure behind the moving contact. The drop in pressure behind the moving contact at the "B" is produced as the anode root moves away from the contact region.

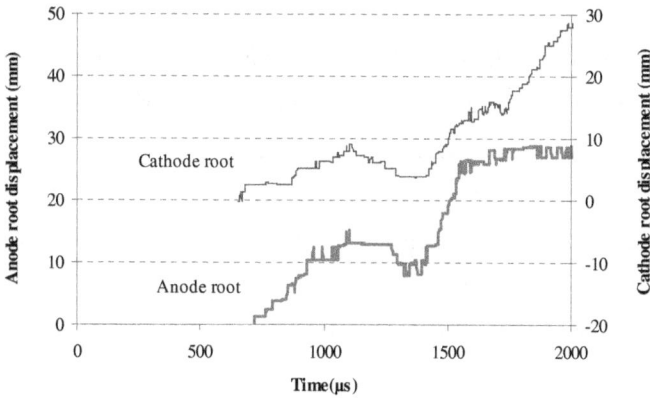

Figure 4.15: Anode root and cathode root displacement of Cu punch contact material, 10 m/s, 2000 A. peak current, choked, ceramic arc chamber

The waveform of anode and cathode root displacement is shown in Figure 4.15. The cathode root moves from the fixed contact region at the period of 600 µs. This event occurred coincident with the small slop of the pressure behind the moving contact.

After the cathode root moves off from the fixed contact, the pressure behind the moving contact starts to rise before the pressure behind the arc stack. The anode root moves off from the contact region about 800 µs. At this point, the pressure in the gap behind the moving contact rises up sharply. While the anode root is in the moving contact region, the profile of the pressure behind the arc stack rises up slowly.

While the waveform of the pressure behind the moving contact swings up and down until the arc root has left the contact region. When the arc roots move off from the contact region, the pressure behind the arc stack is higher than the pressure behind the moving contact.

4.4.2 Arc chamber materials

From literature review in Section 2.4.4, the insulating walls have a significant influence on the arc immobility. The ablated surface affects the characteristics of the plasma, increasing the temperature and gas flow in the arc chamber.

However, it is not completely clear if the arc chamber material affects the pressure in the arc chamber. In addition, at the period at which the arc root moves from the contact region how the pressure in the contact region builds up in the arc chamber.

The effects of the arc chamber materials influence on the pressure build up in the arc chamber are considered. The anode and cathode root displacement are recorded. This is used to find the relationship between the pressure and the arc root at the period at which the arc root moves off from the contact region.

These experiments investigate two types of the arc chamber materials: ceramic and polycarbonate. The investigation was done under the condition of the anode on the moving contact and the cathode on the fixed contact. The summary of the experimental condition is shown in table 4.4.

Experimental Factors	Fixed level
Contact velocity (m/s)	10 m/s
Supply polarity	- Anode on the moving contact - Cathode on the fixed contact
Short circuit current (A)	2000 A
Contact material	Ag/C flat on the fixed contact
Arc chamber venting	Choked
Arc chamber material	Ceramic/Polycarbonate
Gap behind the moving contact	Closed

Table 4.4: Experimental conditions to study the effects of arc chamber materials

The experimental result of the pressure in the fixed contact area is shown in Figure 4.16. The arc root displacement for Polycarbonate arc chamber is shown in Figure 4.17.

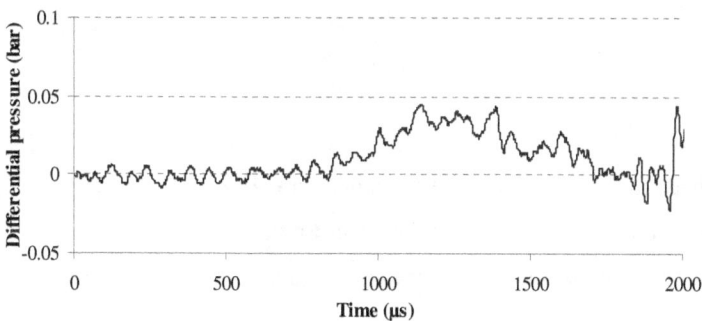

Figure 4.16: Pressure (in the fixed contact region) of Polycarbonate arc chamber

Figure 4.16 presents the waveform of the pressure at the fixed contact region with contact opening velocity 10 m/s in polycarbonate arc chamber. The profile of the pressure stays constant before starting to rise up slightly at the time 800 μs.

The pressure reaches at the peak maximum at the time 1200 μs before the pressure starts to fall down. The profile of pressure declines slightly until the pressure returns to the base level at the period about 2000 μs. The slope of pressure rising up is higher than the slop of decline.

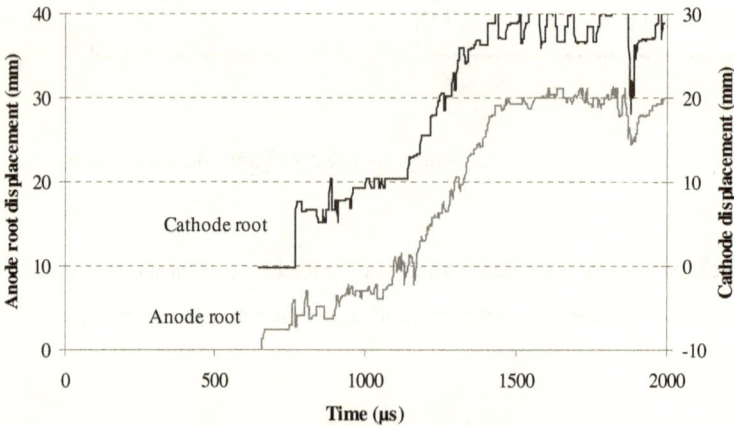

Figure 4.17: Anode and cathode root displacement in Polycarbonate arc chamber

The anode and cathode root displacement is shown in Figure 4.17. When the pressure starts to rise up, the cathode root moves from the contact region. During the period of the pressure climb up, the anode root stays in the contact area and the cathode root moves towards into the arc chamber.

At the time that anode root moves from the moving contact, it is coincident with the maximum pressure. Afterward the pressure starts to drop down and both anode and cathode root move towards into the arc chamber.

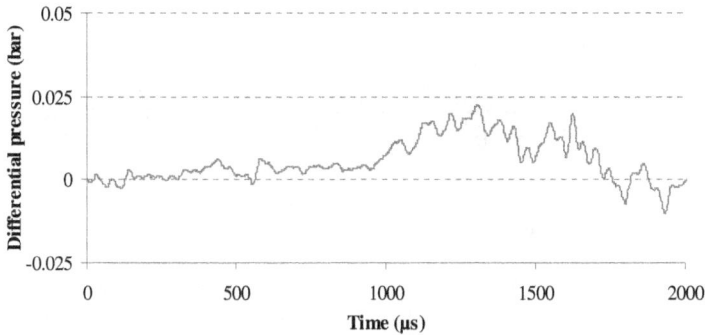

Figure 4.18: Differential pressure (in the fixed contact region) of ceramic arc chamber

The profile of the pressure measurement in the fixed contact region in the ceramic arc chamber is shown in Figure 4.18. At the maximum point, the anode root moves off from the moving contact and the cathode root moves towards into the arc. The profile of the pressure starts to drop down slightly as the arc arrives at the arc stack.

The waveform of the pressure starts to rise up at the period at 900 µs. The maximum pressure is about 0.025 bars at the time 1200 µs. Afterward, the pressure starts to decline slightly down to the original level at the time at 1800µs.

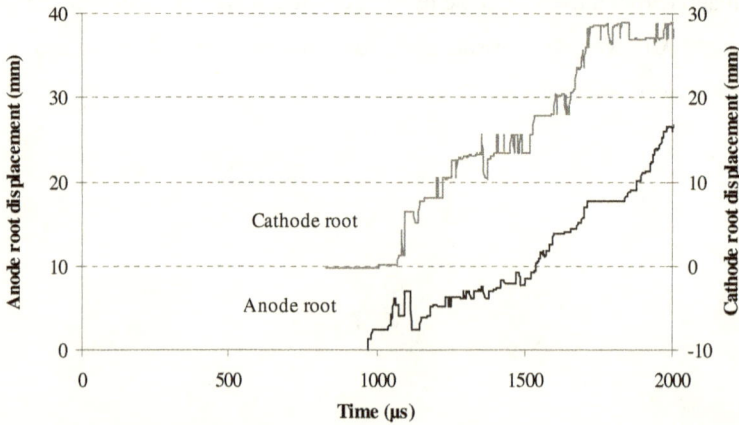

Figure 4.19: Anode root and cathode root displacement of ceramic arc chamber

The anode and cathode root displacements are shown in Figure 4.19. At the point at which the pressure starts to rise up, the anode root still stays in the moving contact region, while the cathode root starts to move from the contact area. After the cathode root departs from the fixed contact area, the pressure increases rapidly.

It reaches the maximum point at the period about 1200µs. At this point, the anode root moves off from the moving contact region and the cathode root moves towards into the arc stack. Afterwards, the pressure drops down slightly from the maximum to the base level at the period of 1800µs.

The pressure in a polycarbonate arc chamber starts to increase before the pressure in a ceramic arc chamber. The maximum pressure in the polycarbonate arc chamber is double that in the ceramic arc chamber.

The arc root in the polycarbonate arc chamber moves from the contact region earlier than the ceramic arc chamber. The arc root contact time in the polycarbonate arc chamber is shorter than the ceramic arc chamber.

4.4.3 Contact opening velocity

From the literature review in Section 2.4.4, Chapter 2, when the contact opening velocity is below 2.2 m/s, the arc root contact time decreases as the contact gap is increased. This is thought to be because of the increase in gas pressure when the contact gap is decreased.

The effects of gas flow on the arc motion from literature review in Section 2.5.2, the modelling shows an important influence on the arc dynamics during an arc moves from the contact region. Increased flow speeds in the re-circulation on the arc chamber produce pressure in the arc chamber.

The experimental results of the arc root contact time in Section 4.3.4, show that the contact opening velocity has an effect on the movement of the arc root from the contact region, especially on the moving contact.

The pressure characteristics in the arc chamber when the arc root starts to move off from contact region with variable contact opening velocity are considered. The influences of the contact opening velocity of 1 m/s on the pressure are inspected. These experiments are used to compare with the experimental results of contact opening velocity of 10 m/s, Polycarbonate arc chamber in the Section 4.4.2, Figure 4.16 and Figure 4.17.

The experimental condition is the anode on the moving contact and the cathode on the fixed contact. The choked arc chamber venting and polycarbonate arc chamber selected. The Ag/C flat is used to observe the events.

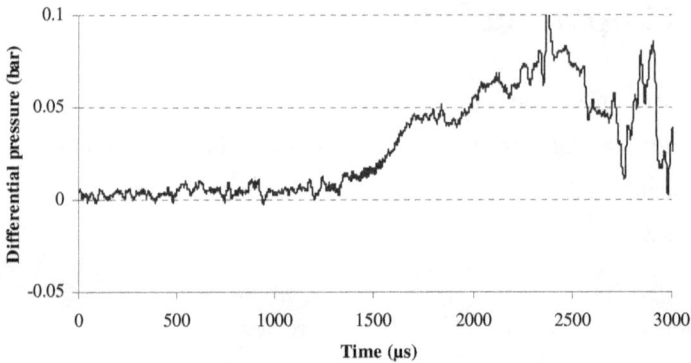

Figure 4.20: The pressure (in the fixed contact region) of Ag/C flat contact material, 1 m/s, Anode on moving contact, 2000 A. peak current, choked, Polycarbonate arc chamber

Figure 4.20 presents the pressure with the contact opening velocity 1 m/s. These results are compared to Figure 4.16 at 10 m/s. The pressure starts to rise up at the period of 1300 µs. The pressure rises up rapidly until reaches at 1800µs before stays constant for a while.

Afterward the pressure second rises up again at 2200 µs before the arc root arrives the peak about 0.1 bar. At the point 2500 µs, the pressure curve starts to drop down shapely.

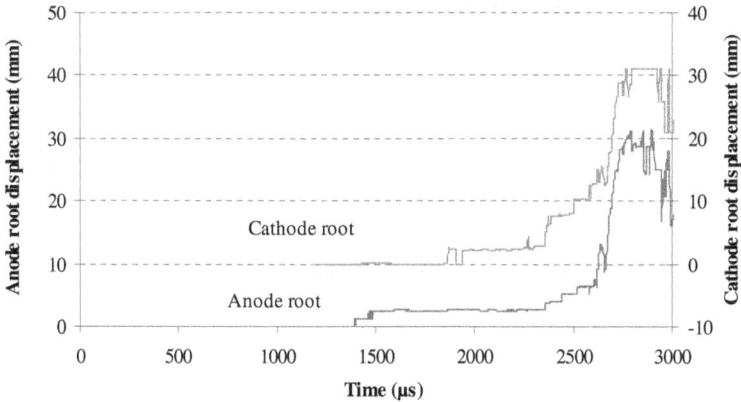

Figure 4.21: Anode root and cathode root displacement of Ag/C flat contact material, 1 m/s, Anode on moving contact, 2000 A. peak current, choked, Polycarbonate arc chamber

The anode and cathode root displacement are shown in Figure 4.21. The results show that when the pressure starts to rise up, both anode and cathode root still stay in the contact region. At the period of 1800 µs, the second pressure rises up.

This is coincident with the point that the cathode root moves off from the fixed contact. When the anode root moves from the contact region at the period of 2600 µs, the pressure drops down shapely.

At contact opening velocity of 10 m/s (see Figure 4.16), the pressure rises earlier than that of 1 m/s. The maximum pressure when the contact opening velocity of 1 m/s is double that of 10 m/s. The arc root contact time at contact opening velocity of 1 m/s is longer than that of 10 m/s. The period the arc root is in the arc chamber at contact opening velocity of 1 m/s is shorter than that of 10 m/s.

4.4.4 Summary of the pressure in the arc chamber

The influences of the gap behind the moving contact on the arc root contact time and pressure in the arc chamber have been investigated. The pressure behind the moving contact rises earlier than that behind the arc stack. After the arc roots move off from the contact region, the pressure behind the arc stack is higher than that behind the moving contact.

The investigations of the arc root motion and the pressure under the effects of the arc chamber materials: polycarbonate and ceramic. In the polycarbonate arc chamber, the pressure increases earlier than the ceramic.

The maximum pressure in a polycarbonate arc chamber is higher than that of a ceramic. The arc roots in the polycarbonate arc chamber move from the contact region earlier than in the ceramic.

The pressure at the contact opening velocity of 10 m/s rises up earlier than that of 1 m/s. However, the maximum pressure of the contact opening velocity of 1 m/s is double higher than that of 10 m/s. The arc root contact times at contact opening velocity of 1 m/s are longer than that of 10 m/s.

4.5 Arc spectrum

From literature review Section 2.4.5, Chapter 2, the Ag I (atom) spectrum is detected in both metallic and gaseous phase. In the metallic phase, the density of metallic vapour decreases due to the increase in gap between electrodes. For Ag/C contact, the carbon reacts with oxygen and forms a deposition. This affects the motion of the arc from the contact region and the dielectric strength.

From the experimental results in Section 4.3, the arc chamber venting, short circuit current level, contact opening velocity and the gap behind the moving contact have a significant effects on the arc root motion. Furthermore, the experimental results from Section 4.4, show that the arc chamber material, gap behind the moving contact and contact opening velocity influence the pressure in the arc chamber.

4.5.1 Arc chamber materials

From literature review in Section 2.4.4, Chapter 2, the gassing walls show a strong effect on the arc root movement. The ablated surface influences the characteristics of the plasma, increasing in temperature, electron density and electric field but decreasing the arc extinction time.

Additionally, the experimental results from Section 4.4, show that the polycarbonate arc chamber affects the pressure in the arc chamber and arc root motion.
The study of the arc spectrum here considers the chemical elements in the arc chamber.

These experiments investigate the arc spectrum of the polycarbonate and ceramic arc chamber under the same condition as in Section 4.4.2.

Figure 4.22 shows spectra lines of polycarbonate arc chamber and table 4.5 shows the possible element species of the polycarbonate arc chamber. Figure 4.23 shows spectra lines of ceramic arc chamber and table 4.6 shows the possible element species of the ceramic arc chamber.

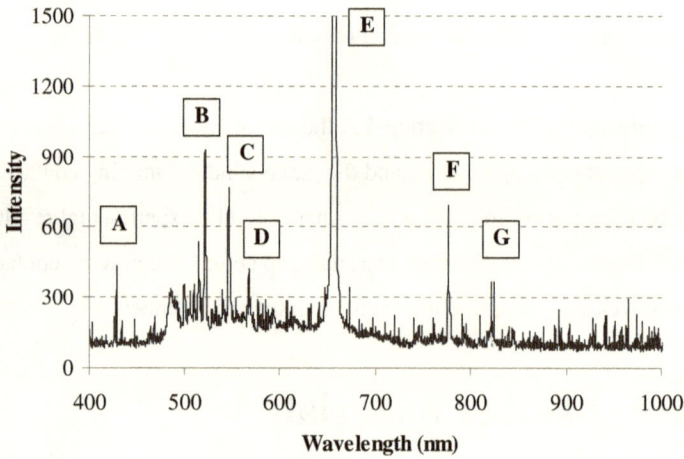

Figure 4.22: Spectra lines of polycarbonate arc chamber, 10 m/s, Anode on moving contact, 2000 A peak current, Ag/C flat, choked arc chamber vent

Spectrum	Wavelength (nm)	Possible element species
A	428	NII, OII
B	515	CV, OII, OIV, NV
C	521, 546	Ag I
D	567	CIII and OV
E	656	NII, H, OII
F	776	OI, NII, CII, CuII
G	821	NI, OI, OII

Table 4.5: Spectra lines and possible element species of polycarbonate arc chamber

Figure 4.22 shows the experimental results of the arc spectrum in the polycarbonate arc chamber. The results show nine spectra lines occurred in the arc chamber. There are 428, 515,521, 546, 567, 656, 776, 821 and 824 nm. The possibility element species are shown in table 4.5. The maximum intensity is the spectra at the wavelength E.

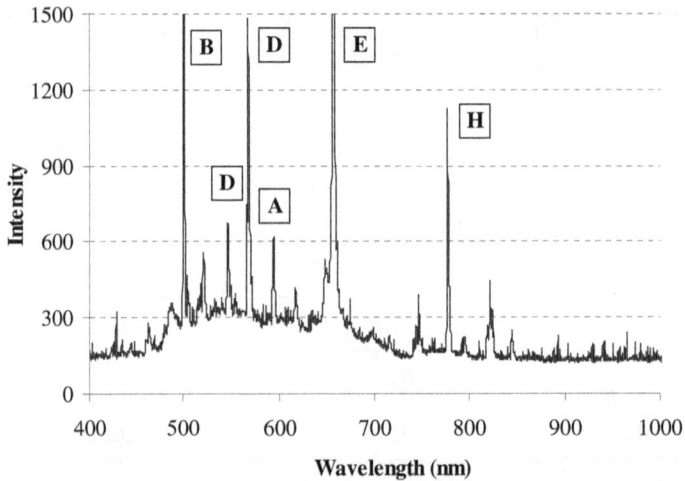

Figure 4.23: Spectra lines of ceramic arc chamber10 m/s, Anode on moving contact, 2000 A peak current, Ag/C flat

Spectrum	Wavelength (nm)	Possible element species
B	521	Ag I
C	546	Ag I
D	567	CIII and OV
E	656	NII, H, OII
F	776	OI, NII, CII, CuII
G	821	NI, OI, OII
H	500	OII, OIV, TIVIII, AlIV, NIII
I	594	OIV, CII, AlIV, NIV
J	746	NI, NII, OIII

Table 4.6: Spectrum lines and possible element species of ceramic arc chamber

The spectra lines in the ceramic are chamber are shown in Figure 4.23. The experimental results show nine spectra lines at the wavelength 521, 546, 567, 656, 776, 821, 500, 594 and 746 nm. The possibility elements species are shown in Table 4.6. At the wavelength B, D and E show clearly with a high intensity spectrum.

The spectrum lines show both in the polycarbonate and ceramic arc chamber are the spectrum at the wavelength B, C, D, E, F, G. The elements of NII and OII at the wavelength 428 nm is shown only in the polycarbonate are chamber.

On the other hand, the spectrums at the wavelength 500 and 594 nm are shown only in the ceramic arc chamber. The average intensity of the arc spectrum in the ceramic arc chamber is higher than in the polycarbonate.

4.5.2 Contact velocity

From experimental results in Section 4.2.3 and Section 4.4.3, the contact opening velocity has effects on the arc root motion and pressure in the arc chamber. From the literature review in Section 2.4.5, the molecules and ions of the surroundings gases contribute to discharge.

The density of metallic vapour decreases the gap between electrodes increases. The arc spectrum from the optical spectrometer can indicate the possible element species occurring at different contact opening velocities.

The effects of the contact opening velocity on the elements species in the arc chamber are observed. Two speeds of contact opening velocity: 10 m/s and 1 m/s are used to inspect the events. Most of the experimental condition in this experiment is similar to the condition in the Section 4.4.2.

The arc spectrum with a contact opening velocity of 1 m/s is shown in Figure 24.
Table 4.7 shows the possible element species with contact opening velocity of 1 m/s.

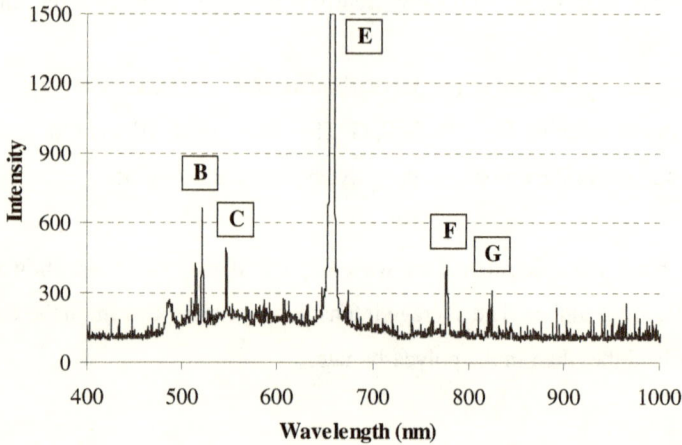

Figure 4.24: Spectra lines of contact velocity 1 m/s, polycarbonate, Ag/C flat, choked arc chamber

Spectrum	Wavelength (nm)	Possible element species
B	515	CV, OII, OIV, NV
C	521, 546	Ag I
E	656	NII, H, OII
F	776	OI, NII, CII, CuII
G	821,824	NI, OI, OII

Table 4.7: Spectra lines and possible element species of contact velocity 1 m/s

Figure 4.24 shows the experimental results of the arc spectrum with the contact opening velocity 1 m/s. The experimental results show seven peaks in the spectrum. There are at the wavelength 515, 521, 546, 657, 776, 821 and 824 nm. The possible element species are shown in table 4.7.

All of these spectra are similar to the experimental results with the contact opening velocity 10 m/s, see Figure 4.22, Section 4.5.1. The different peaks are the wavelength 428 and 567 nm, which show only in the contact opening velocity of 10 m/s. The maximum intensity shows at the wavelength 657 nm.

The wavelength of the spectra at C, the elements Ag I occurs on both 10 and 1 m/s. The average intensity of spectra line at the contact opening velocity 10 m/s is clearer than that of 1 m/s.

4.5.3 Contact materials

From literature review in Section 2.4.1, Chapter 2, the contact material affect the arc root mobility. The arc root on Cu contact could move by the self magnetic field of the Cu conductor. From literature review in section 2.4.5, Chapter 2, a charge-coupled device (CCD) can detect the element species of Ag I emitted from the silver contact material.

There are numerous experiments and modelling on the influences of the contact material. However, there are few experiments that observe the element species of the contact materials by the optical spectrometer. Therefore, the arc spectrum and element species of contact materials: Cu punch and Ag/C step are investigated. The experimental condition is shown in Table 4.8.

Experimental Factors	Fixed level
Contact velocity (m/s)	10 m/s
Supply polarity	- Anode on the moving contact - Cathode on the fixed contact
Short circuit current (A)	2000 A
Contact material	Ag/C step and Cu punch
Arc chamber venting	Choked
Arc chamber material	Ceramic
Gap behind the moving contact	Opened

Table 4.8: Experimental conditions to study the effects of contact materials

Figure 4.25 shows the spectra lines of Ag/C step contact material and table 4.9 shows the possible element species of the Ag/C step contact material. Figure 4.26 shows the spectra lines of Cu punch contact material and table 4.10 shows the possible element species of the Cu punch contact material.

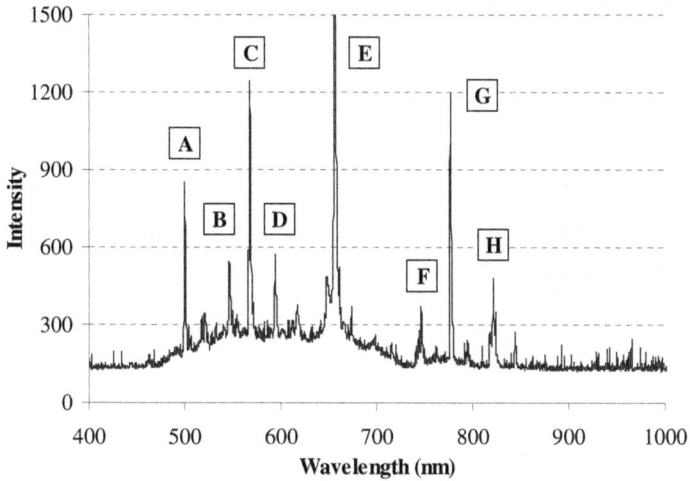

Figure 4.25: Spectra lines of Ag/C step contact material, 10 m/s, Anode on moving contact, 2000 A. peak current, choked, Ceramic arc chamber

Spectrum	Wavelength (nm)	Possible element species
A	500	OII, OIV, TIVIII, AlIV, NIII
B	546	AgI
C	567	CIII and OV
D	594	OIV, CII, AlIV, NIV
E	656	NII, H, OII
F	746	NI, NII, OIII
G	776	OI, NII, CII, CuII
H	821	NI, OI, OII

Table 4.9: Spectra lines and possible element species of Ag/C step contact material

Figure 4.25 shows the spectrum lines of contact material Ag/C step. The experimental results show eight spectra lines happened in the arc chamber. They are spectrum at the wavelength 500, 546, 567, 594, 656, 746, 776 and 821 nm.

The possibility of the elements is shown in table 4.9. The spectrum at the wavelength 656 nm shows the maximum intensity. The minimum intensity of the arc spectra is at the wavelength at 746 nm.

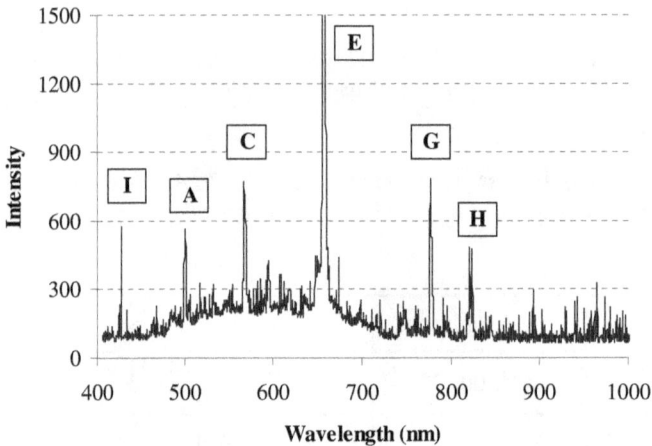

Figure 4.26: Spectra lines of Cu punch contact material, 10 m/s, Anode on moving contact, 2000 A. peak current, choked, Ceramic arc chamber

Spectrum	Wavelength (nm)	Possible element species
I	428	NII, OII
A	500	OII, OIV, TIVIII, AlIV, NIII
C	567	CIII and OV
E	656	NII, H, OII
G	776	OI, NII, CII, CuII
H	821	NI, OI, OII

Table 4.10: Spectra lines and possible element species of Cu punch contact material

The arc spectrums of Cu punch contact material are shown in Figure 4.26. Seven spectra lines are occurred in the arc chamber. They are the spectra at the wavelength 428, 500, 567, 656, 776, 821 and 824 nm. The possibility elements are shown in Table 4.10. The maximum intensity is the spectrum 656 nm. These elements species could dominate the arc root mobility.

The spectrums occur on both the Ag/C step and Cu punch are at the wavelength at B, D, and F. The spectra line occur only with the Cu punch contact material, is the wavelength at 428 nm. Only in the Ag/C step contact material shows the element species of AgI at the wavelength 546 nm.

There are five spectrum lines that appear in both Cu punch and Ag/C step as shown at the wavelength 500, 567, 656, 776 and 821 nm. The intensity of Ag/C step spectrum is greater than Cu punch.

4.5.4 Summary of spectrum analysis

The spectrum at 428 nm shows only in the polycarbonate arc chamber. The spectrums at 500 nm and 594 nm show only in the ceramic arc chamber. The spectrums at the wavelength 428 and 567 nm show only with the contact opening velocity of 10 m/s. The spectrum line at 428 nm shows only with Cu punch while Ag I at 546 nm shows only with Ag/C contact.

The spectrum at the wavelength 428 nm shows under the condition of the Cu punch contact material and with contact opening velocity of 10 m/s. The probable species of this spectrum are nitrogen and oxygen. Under these conditions, the element species may come from the chemical process since these element species are the main components of air.

The species of aluminium show at the spectrum wavelength 500 and 594 nm. This element is a component of the ceramic arc chamber. It shows that these elements emitted from the arc chamber.

4.6 Summary

The Ag/C on the moving contact is not suitable for commercial use in miniature circuit breakers. The influence of the arc chamber venting has effects on the arc root on the moving contact. The arc root contact times increase as the vent area is decreased. The arc root contact time decrease as the short circuit current is increased.

The contact opening velocity has a significant influence on the mobility of the arc root moving off from the contact region. The arc root contact time with the gap opened is longer than with the gap closed.

The gap behind the moving contact has influence on the arc root and pressure in the arc chamber. The maximum pressure in the polycarbonate arc chamber is double higher than the ceramic. The maximum pressure of the contact opening velocity of 1 m/s is double higher than that of 10 m/s.

The arc spectrum at 428 nm shows in the polycarbonate arc chamber while the spectrums at 500nm and 594 nm only in the ceramic. The spectrum line at 428 nm shows in Cu punch while Ag I at 546 nm shows in Ag/C contact.

CHAPTER 5

MODELLING RESULTS

5.1 Introduction

In this chapter the magnetic forces, gas dynamic forces and thermal energy are modelled to determine the influence contact velocity on the arc root motion from the contact region. Experimental data from pressure measurement are used to investigate the gas flow in the arc chamber. The optical data are used to analysis arc root velocity from the contact region. The mass flow in the contact region is obtain from the thermal energy.

5.2 The modelling of the magnetic forces in the contact area

The model considered in this book was very simple and uncomplicated. The conductors in circuit breaker and the arc are assumed to be threadlike. The anode is set on the moving contact and the cathode is on the fixed contact. The magnetic flux density (B) is calculated by using a function of current and contact gap.

The experimental results provide the data of the arc current and the contact gap. The magnetic flux density (B) is obtained from the magnetic force modeling. The magnetic forces on arc root are calculated by using the relationship of the arc current, contact gap and magnetic flux density (B).

At the point at which the arc root moves from contact region with contact opening velocity 1-10 m/s, the contact gap is between 1.57 to 6 mm. Thus, the arc motion for a contact gap less than 1 cm can be explained in terms of magnetic field [3,21,70].

The conductors carrying a current (i), create the magnetic field B. The electromagnetic force acting on the arc is given by $\vec{F}_{mag} = \vec{B}.i.a$ as a simplified version. The definition of "a" is the length of the arc as shown in Figure 5.1.

The magnetic flux density (B) was computed by using a function of current and contact gap. The boundary of modelling on the cathode root is limited to the length of the fixed contact region, about 3 mm. The boundary of the anode root is limited to the edge of the moving contact region, about 10 mm.

When the arc root stays in the contact region, it can be considered as a short arc and the arc structure has only anode and cathode root [2]. Since the real arc root lengths (cathode sheath and anode sheath) could be as low as 10^{-2} mm [23,33].

Therefore, the center of the arc root about 0.5×10^{-2} mm from the electrode surface, is used to simulate the magnetic force on the arc root. This model is implemented in MATLAB software for magnetic forces calculation and analysis. There are two cases as followings:

Case I: Simplified model

It is assumed that the current flows completely in one line and in the surface of the conductors. The distribution of current density in the conductor is ignored. The contact materials or contact geometries are assumed to have a minimal effect as shown in Figure 5.1.

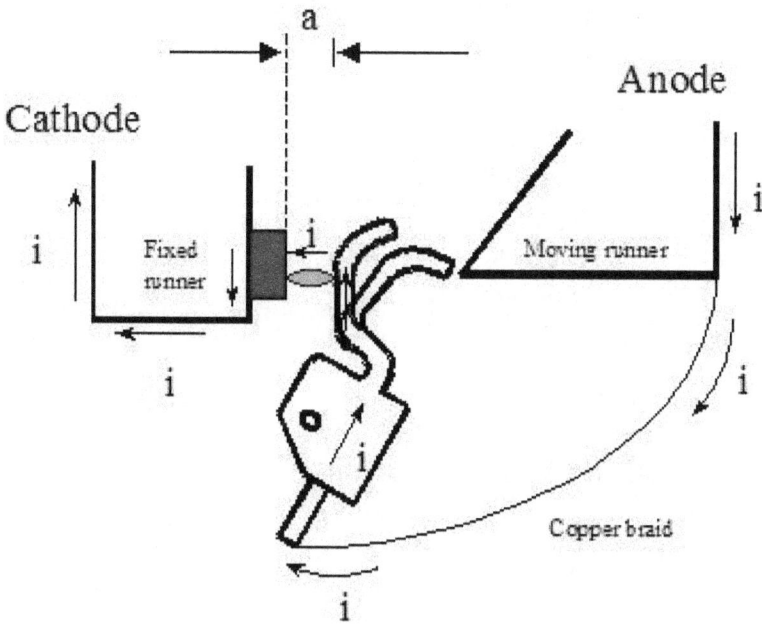

Figure 5.1: Current limited circuit diagram with two contact positions

The anode is set on the moving contact and the cathode is on the fixed contact as element $\overline{BG}$ and $\overline{FC}$ as shown in Figure 5.2. The arc is laid on the element $\overline{GF}$ at the coordinate (X_f, Y_f, Z_f) and (X_g, Y_g, Z_g).

The current originates from the point B (X_b, Y_b, Z_b), passed through G (X_g, Y_g, Z_g) and F (X_f, Y_f, Z_f). The current ends at the point C (X_c, Y_c, Z_c). The magnetic field is calculated at S (X_s, Y_s, Z_s).

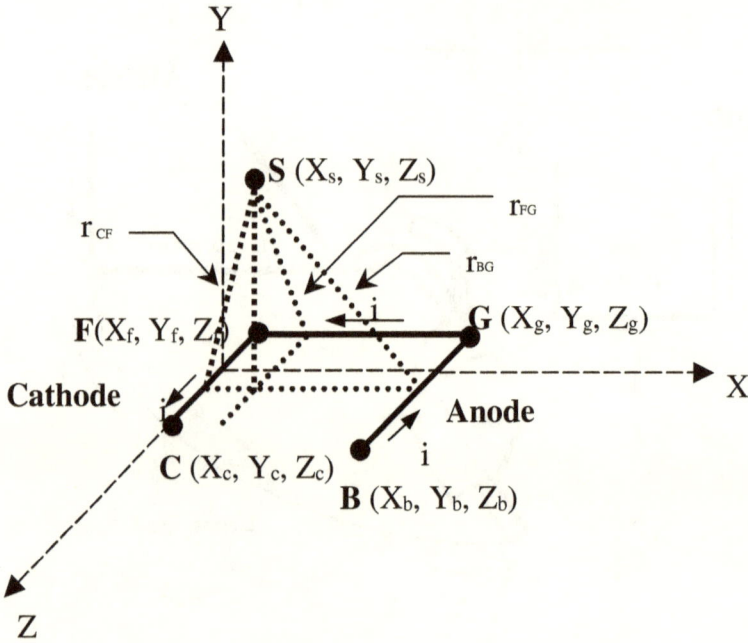

Figure 5.2: Configuration of the arc ($\overrightarrow{GF}$) in the contact region

The magnetic potential can be determined at any point "S" of the contact region [48] by the equivalent circuit shown in Figure 5.2.

$$\overrightarrow{A_{GF}} = \overrightarrow{a_x} \frac{\mu_o \mu_r}{4\pi} i \int_{G}^{F} \frac{dx}{(r^2 + x^2)}$$

Equation 5-1

$$\overrightarrow{A_{BG}} = -\overrightarrow{a_z} \frac{\mu_o \mu_r}{4\pi} i \int_{B}^{G} \frac{dz}{(r^2 + z^2)}$$

Equation 5-2

$$\overrightarrow{A_{FC}} = \overrightarrow{a_z} \frac{\mu_o \mu_r}{4\pi} i \int_F^C \frac{dz}{(r^2 + z^2)}$$

Equation 5-3

a_x, a_z : Vector fields in x and z direction

μ_o : Permeability of free space = 4π x 10^{-7} Webers/m-amp

μ_r : Relative permeability =1.0

r : Magnitude of distance from current moment to point S at
    which magnetic effect is evaluated, (meter)

Magnetic potential A_x in air is caused by current from G to F, ($A_{GF} = A_x$). Magnetic potential A_z in air is caused by current from B to G, ($A_{BG} = - A_z$) and F to C, ($A_{FC} = A_z$). The magnetic flux density ($\vec{B}$) was obtained from the curl of magnetic potential $\vec{A}$ as:

$$\vec{B} = \vec{\nabla} \times \vec{A}$$

Equation 5-4

$$\vec{B} = (\frac{\partial A_z}{\partial y} - 0)\overrightarrow{a_x} + (\frac{\partial A_x}{\partial z} - \frac{\partial A_z}{\partial x})\overrightarrow{a_y} + (0 - \frac{\partial A_x}{\partial y})\overrightarrow{a_z}$$

Equation 5-5

Therefore, the magnetic flux density $\vec{B}$ is defined by three components, $\vec{B}_x, \vec{B}_y, \vec{B}_z$.

$$B_x = \frac{\partial Az}{\partial y}$$

Equation 5-6

$$B_y = \frac{\partial A_x}{\partial z} - \frac{\partial A_z}{\partial x}$$

Equation 5-7

$$B_z = -\frac{\partial A_x}{\partial y}$$
Equation 5-8

These three elements are similar to the components obtained by means of Biot - Savart's law [20,21,22, 58].

$$\vec{B} = \frac{\mu_o}{4\pi} \int \frac{I\vec{dl} \times \vec{a_r}}{r^2}$$
Equation 5-9

The modeling results for case I, the magnetic forces per unit length as a function of contact gap and arc current is shown in Figure 5.3.

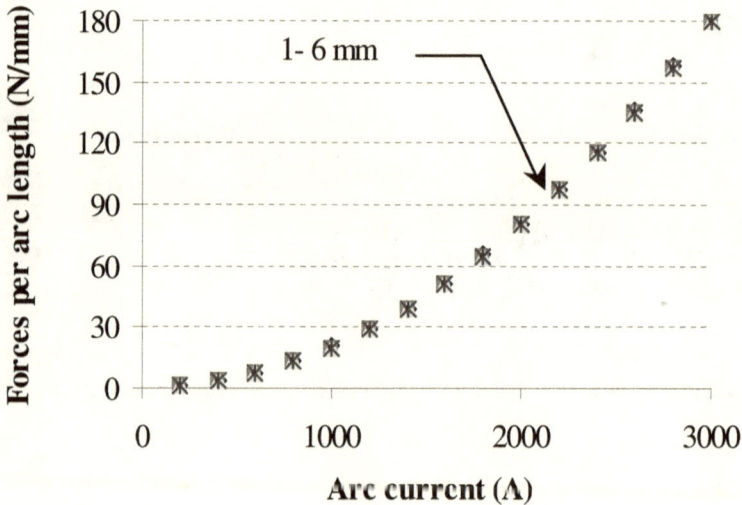

Figure 5.3: Forces per arc length with short circuit arc current and contact gap

The modelling results in figure 5.5 show that the magnetic forces on the arc root increase as the arc current increases. The contact gap has a minimal influence on the magnetic forces on the arc root. The magnetic forces on the arc root are mainly dependent on arc current

Case II: Model including current distribution and step

This modeling is based on case I. The different is the influence of the shapes of the contacts are modelled. From the experimental results in Section 4.3.1, Chapter 4, the Ag/C on the moving contact causes a longer delay in the tip of the moving contact. An Ag/C step and Ag/C flat on the fixed contact are used to compute the magnetic forces on the arc root, see Figure 3.19, Chapter 3.

The current distribution inside the conductors with the presence of the arc are shown in Figure 5.3 and 5.4. It is assumed that the conductor is divided into 20 partial parts and 21 current elements each carrying equal current separately by 5×10^{-2} mm.

The material of the arc runner is Cu and it is 1.0 mm thick. The material of the contact is Ag/C (95/5 %) and 0.8 mm thick. The anode is on the moving contact and the cathode is on the fixed contact.

A Silver/Graphite contact pad with a step to the surface of the arc runner is shown in Figure 5.4. A Silver/Graphite contact pad flushes without a step to the surface of the arc runner is shown in Figure 5.5.

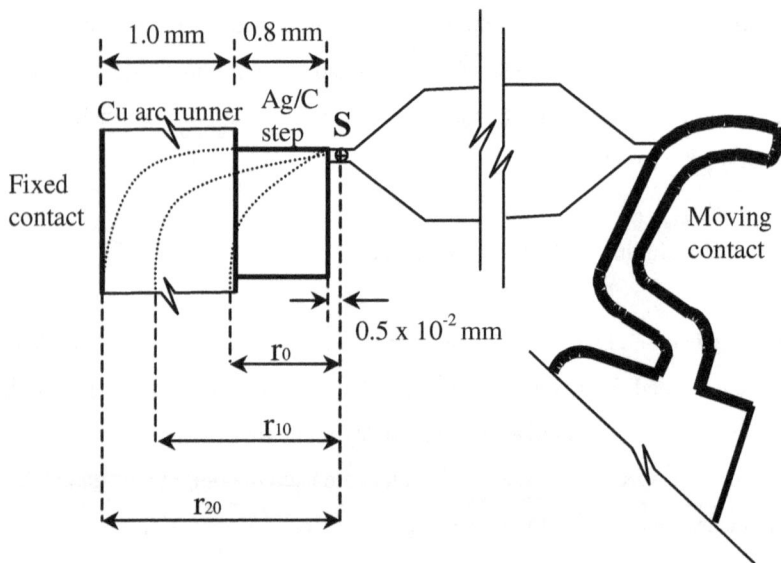

$$r_0 = 80.5 \times 10^{-2} \text{ mm.}$$

$$r_1 = 85.5 \times 10^{-2} \text{ mm.}$$

.

.

$$r_{20} = 180.5 \times 10^{-2} \text{ mm.}$$

Figure 5.4: Current distribution in fixed contact Ag/C step with arc root

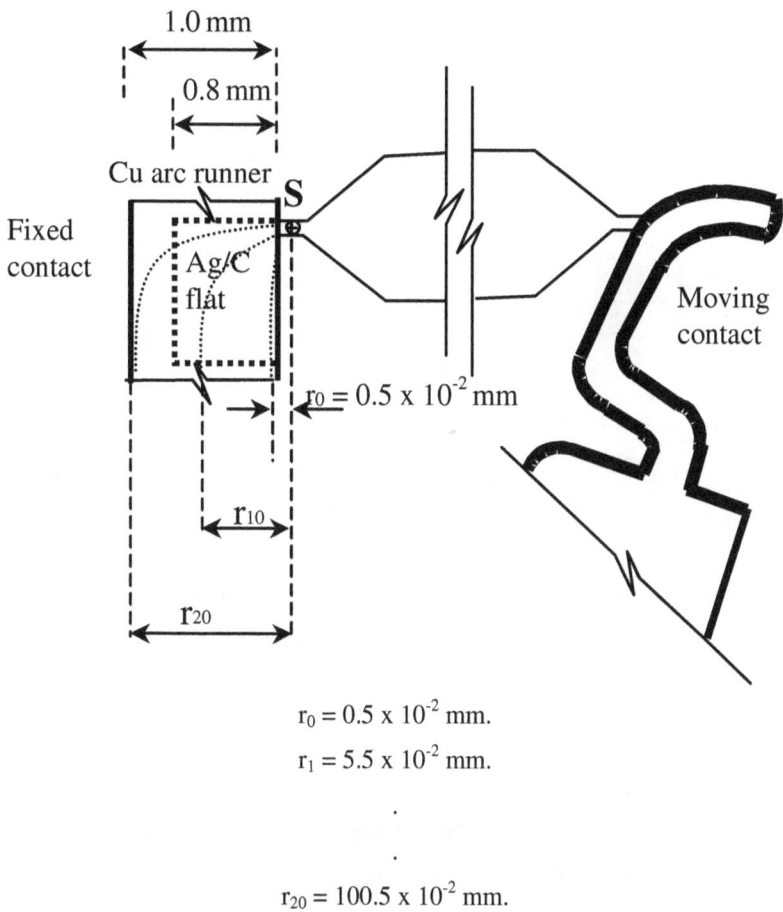

$r_0 = 0.5 \times 10^{-2}$ mm.

$r_1 = 5.5 \times 10^{-2}$ mm.

.

.

$r_{20} = 100.5 \times 10^{-2}$ mm.

Figure 5.5: Current distribution in fixed contact Ag/C flat with arc root

The magnetic forces modeling for case II is shown in Figure 5.6.

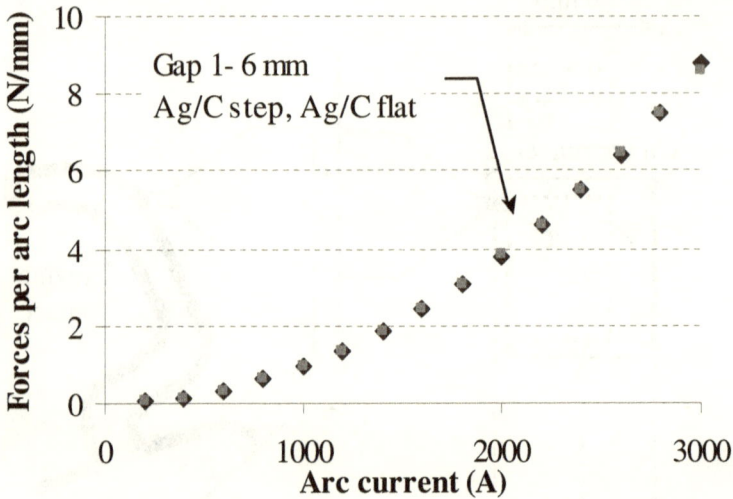

Figure 5.6: Force per arc length on Ag/C flat and Ag/C step with contact gap

The modelling results in figure 5.6 show that the magnetic forces on the arc root increase as the arc current increases. The contact gap has a minimal influence on the magnetic forces on the arc root. The magnetic forces on the arc root are mainly dependent on arc current. Both Ag/C flat and step have no significant influence of the contact gap on the magnetic forces. The magnetic forces are very high compared to [11,31].

The magnetic forces for case II is much lower than case I. This could be due to the assumption in case I the current completely flows in one line and only on the surface of conductors. In case II the current distribution is divided into 21 parts and laid along the thickness of the conductor.

These models show that the current distribution in the conductor has an significant effect on the magnetic forces when the arc root starts to move from the contact region. The effect of the step in the fixed contact is minimal on the magnetic forces.

However, the current density, type of materials and the thickness of conductors have not taken into account. These factors are a subject for future work including numerical modelling of the current flow in the contact region. For comparison with other researcher only the simplified model is used.

5.3 The modelling of the gas dynamic in the contact area

It is assumed that the arc root moves at the same velocity as the gas flow behind the shock wave. The moving arc causes a displacement of the gas, causing a gas dynamic force, F_g, on the arc root [38,62].

A simple model of the arc assumes that the arc is a weightless piston moving in a narrow channel with a velocity u_1. A shock wave forms in front of the arc. The gas behind the shock front is heated up. This results in high temperature and a raise in pressure [14,47].

For a steady flow the arc moves at the same velocity as the gas behind the shock wave. The flow velocity u_1 of the gas (pressure P_1) behind a shock wave propagating into still air (sound speed = a_o, P_o= pressure in front) is given by:

$$\frac{u_1}{a_o} = (\frac{P_1}{P_o} - 1)(1 - \mu^2) \sqrt{\frac{1}{(\frac{P_1}{P_o} + \mu^2)(1 + \mu^2)}}$$

Equation 5-10

$$F_g = (\frac{P_1}{P_o} - 1) . P_o . d . l$$

Equation 5-11

$$F_g = (P_1 - P_0) . A$$

Equation 5-12

Where A is the area of the channel i.e. the arc chamber in the contact region.

P_1 : the gas pressure behind the shock

P_o : the pressure in front of the shock

$\mu^2 = (\gamma - 1)/(\gamma + 1)$ where $\gamma = 1.14$ (air)

a_o : sound speed $= 331.2$ m/s, at temperature 20 °C

d : arc dimension

l : arc length

The gas dynamic forces are estimated from the different pressure using equation 5-11. Assuming the dimension of the arc is very small. Therefore, the force is dependent upon the difference of the pressure across the arc.

From the gas dynamic modelling, the velocity of the arc flow in the arc chamber is approximated from the ratio of the pressure across the arc, as described in equation 5-10. The results are shown in Figure 5.7.

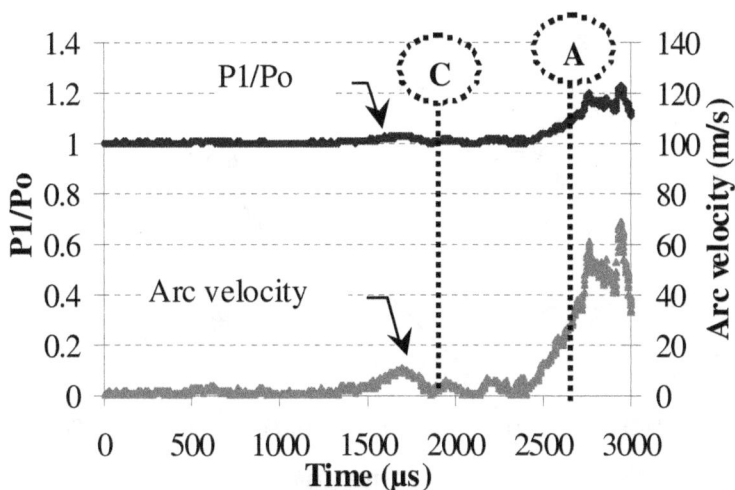

Figure 5.7: Calculated arc flow velocity and pressure across the arc, Point "A" is the point at which the anode root moves from the moving contact and point "C" is the point at which the cathode root moves from the fixed contact, at contact opening velocity 10 m/s and 1 m/s, ceramic arc chamber, Ag/C step

The arc flow velocity increased as the ratio of pressure in the arc chamber increased as shown in Figure 5.7. The arc flow velocity of the cathode moves from the fixed contact region is lower than the anode commutes from the moving contact region. The velocity of the flows increases rapidly after the arc root commutes from the contact region.

The arc root velocity in the arc chamber from the optical data is shown in Figure 5.8 and 5.9.

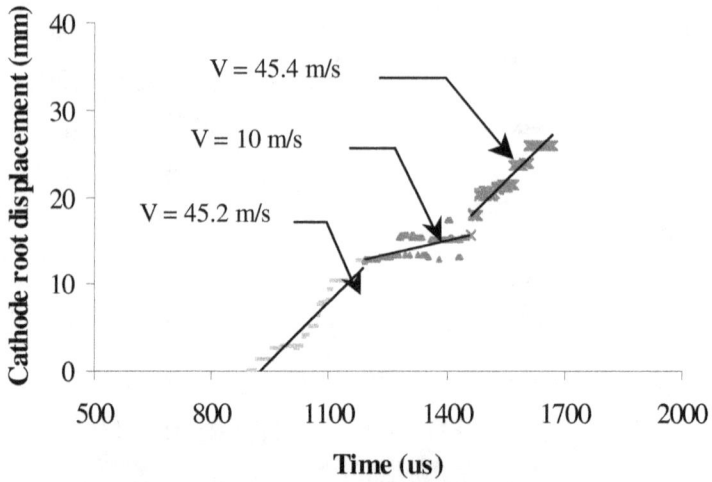

(a) Contact opening velocity 10 m/s

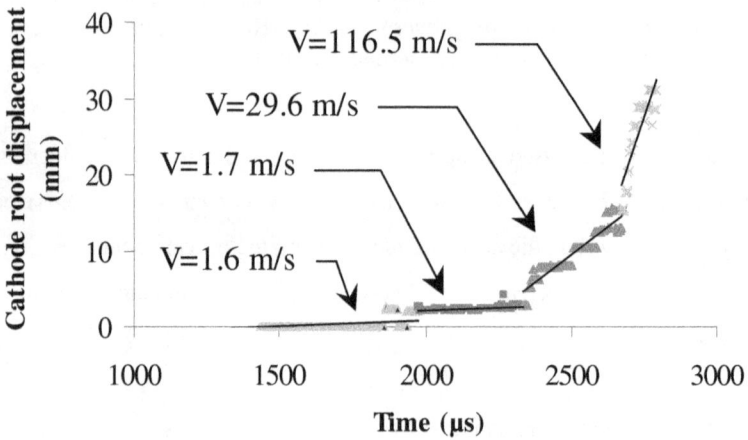

(b) Contact opening velocity 1 m/s

Figure 5.8: Experimental cathode root velocity, at contact velocity 10 m/s and 1 m/s, ceramic arc chamber, Ag/C step

The velocity of the cathode root moves from contact region and moves toward the arc stack is shown in Figure 5.8. The cathode root, at contact opening velocity 10 m/s, moves off from the fixed contact area about 45.2 m/s. At contact opening velocity of 1 m/s, the cathode root moves off with velocity 1.6 m/s.

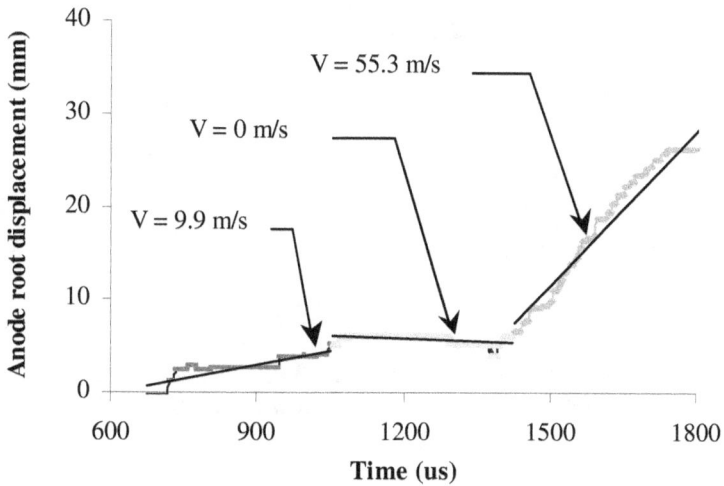

(a) Contact opening velocity 10 m/s

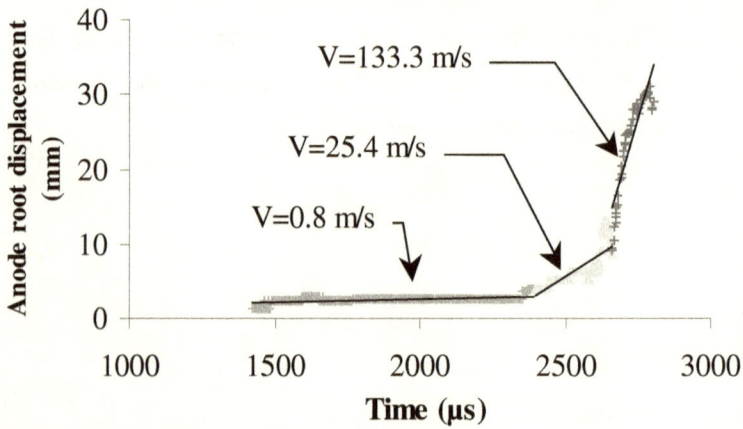

(b) Contact opening velocity 1 m/s

Figure 5.9: Experimental anode root optical displacement and velocity, at contact opening velocity 10 m/s and 1 m/s, ceramic arc chamber, Ag/C step

Figure 5.9 show the anode root velocity with contact opening velocity of 1 m/s. The anode root at contact opening velocity of 10 m/s commutes from the moving contact region with velocity of 55.3 m/s. At contact opening velocity of 1 m/s, the arc root moves off at the speed 25.4 m/s.

From Figure 5.7 and Figure 5.9, at low contact opening velocity of 1 m/s, the velocity of the gas flow from calculation is higher than the arc root from experiment.

5.4 Arc power

It can be assumed that the arc energy is dissipated in the contact area for a short arc. This energy is transmitted to heat the contact and also work for the arc displacement. The relationship between the arc energy, heat and work for the arc root displacement is shown as followings:

$$V \cdot i \cdot t = W + Q \qquad\qquad \text{Equation 5-13}$$

$$P = V \cdot i = \frac{dW}{dt} + \frac{dQ}{dt} \qquad\qquad \text{Equation 5-14}$$

When $W = F \times S$,

F = Force

S = Distance of the arc root movement

Then $\dfrac{dW}{dt} = F \times \dfrac{dS}{dt} = F \times v \qquad\qquad$ Equation 5-15

When v = Arc root velocity

From equation 5-15, the work rate is a function of force and arc velocity. The force is obtained from the magnetic forces modelling as described in Section 5.2. The arc root velocity can be obtained directly from experiments of the optical data and calculation of the gas dynamic modelling. Thus, the work rate can be estimated.

From equation 5-14, the work rate is known and the arc power can be obtained directly from the experimental results of the arc current and voltage. Hence, the heat deposited in the contact area can be approximated.

The typical waveforms of the arc voltage and arc current with contact opening velocities 10 m/s, 5.5 m/s, 4.0 m/s and 1 m/s are show in Figure 5.10 and 5.11. The arc power with different contact opening velocities are shown in Figure 5.12.

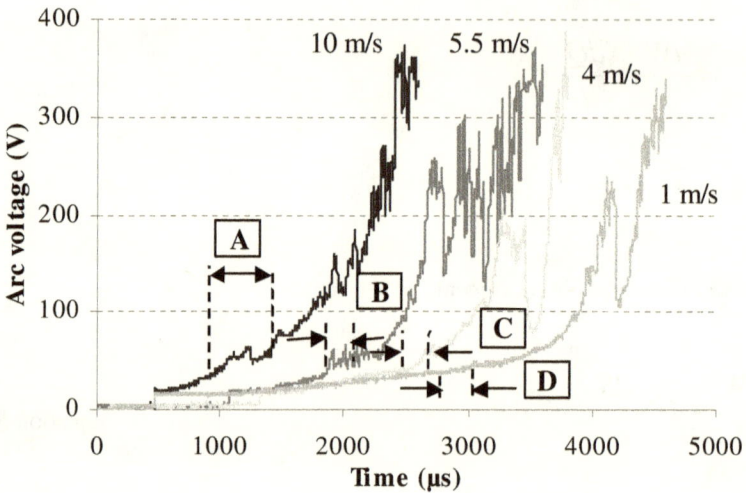

Figure 5.10: Arc voltage and contact opening velocity with contact gap at the point the arc root moves from the contact region, position A, B, C, D are the time periods between the cathode root moves (start) and when the anode root moves (end) from the contact region at contact opening 10.0, 5.5, 4.0 and 1.0 m/s, Ag/C step contact material, Ceramic arc chamber, Choked arc chamber venting, gap behind the moving contact opened (from Section 4.3.4, Chapter 4)

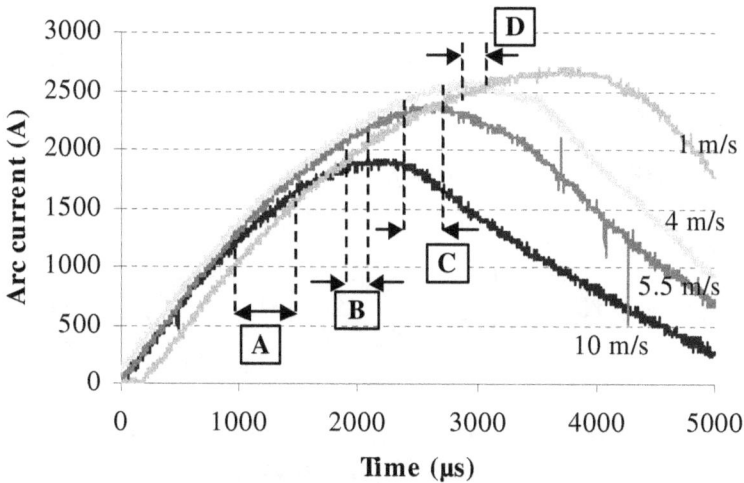

Figure 5.11: Arc current and contact opening velocity with the time that arc root moves from the contact region. Position A, B, C, D are the time periods between the cathode root moves (start) and when the anode root moves (end) from the contact region at contact opening 10.0, 5.5, 4.0 and 1.0 m/s, Ag/C step contact material, Ceramic arc chamber, Choked arc chamber venting, gap behind the moving contact opened(from Section 4.3.4, Chapter 4)

Figure 5.10 shows the arc voltage for different contact opening velocities. The variation in contact gap at the time at which the arc root moves from the contact region is also shown. At 1 m/s opening velocity the rate of increase of arc voltage is lower than at 10 m/s. The contact gap at the time that the arc root starts to move from the contact region decreases as the contact opening velocity increases.

These results are in accord with previous work [23]. Additionally, the delay for arc root commutation is longer at the lower contact opening speed. The experimental results show that the arc current at the point at which the arc root moves from the contact region increased as contact opening velocity decreased as shown in Figure 5.11.

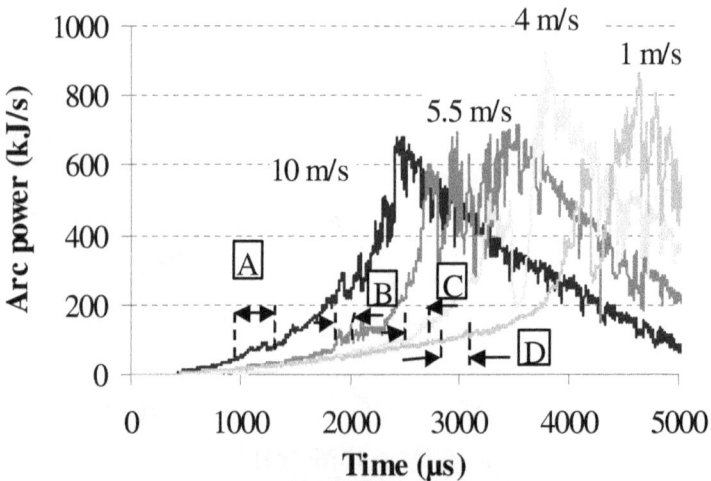

Figure 5.12: Arc power and contact opening velocity with the time that arc root moves from the contact region. Position A, B, C, D are the time periods between the cathode root moves (start) and when the anode root moves (end) from the contact region, region at contact opening 10.0, 5.5, 4.0 and 1.0 m/s, Ag/C step contact material, Ceramic arc chamber, Choked vent, gap behind the moving contact opened(from Section 4.3.4, Chapter 4)

The variation of arc power with contact opening velocity and the point at which the arc root commutes from the contact region are shown in Figure 5.12. There is little variation in arc power with contact opening velocity. The arc power is a function of arc current and arc voltage as well as arc root contact time.

At low contact velocity the arc voltage at the point at which the arc root moves from the contact region is lower than at high contact opening velocity but the arc current is higher. Therefore, the arc power at the point at which the arc root moves from the contact region on the different contact opening velocity are in the same range.

5.5 Thermal energy

For a short arc, most of the arc energy is dissipated in the contact area and much of this energy is transmitted to electrodes. The contact material is heated, melted, vaporized and ionised. The resulting gas expands generating a thermally driven flow through the arc chamber.

It is proposed [23] that a balance between the flow and the Lorentz force $\vec{j} \times \vec{B}$ on the arc is a governing factor in arc immobility. A characteristic contact gap is proposed at the point that the arc root moves from the contact region given by:

$$a = \frac{V_A}{4B} \frac{\rho_s}{h} \sqrt{\gamma RT}$$

Equation 5-16

where   a   = Contact gap
V_A = Arc voltage
B   = Magnetic flux density
ρ_s = Density of contact material
h   = Enthalpy of metal vapour (overheating energies, melting volatilization and ionization)
= Specific heat ratio = C_p/C_v
R   = R_o/mass per mole
R_o = Ideal gases constant
T   = Temperature

A new approach is proposed which uses the energy exchanges in the contact area to calculate the gas flow from the contact region. This can be used with previous theories relating gas dynamics to arc mobility [11] to take account of these thermal effects and to predict arc immunity.

The enthalpy of the metal vapour (h) is a key parameter in the calculation. It is assumed that the Ag/C contact behaves thermally as pure Ag. The total enthalpy of the metal vapour is obtained assuming the following stages.

Stage 1: The metal is heated to melting point

The specific heat c_p of metal at high temperature can be calculated by [71]

Solid state: $\quad C_{ps} = A + BT_m + CT_m^2 + DT_m^{-2}$ $\qquad$ Equation 5-17

From the reference data [11], A = 21.30, B = 8.54x10^{-3}, C=0, D = 1.51x10^5 and the maximum melting temperature is 1233.95 K. The specific heat (c_{ps}) of Ag at the solid state is 295.494 J Kg^{-1} K^{-1}.

It is assumed that the starting point temperature of the metal is equal to environment temperature (293 K). The enthalpy change of the metal from starting point temperature to the melting point is calculated from:

$$\Delta h = c_{ps} \cdot (T_m - T_s)$$ $\qquad$ Equation 5-18

Stage 2: Melting the metal (at 1233.95 K, the enthalpy of fusion is $102.315 \times 10^3 \text{JKg}^{-1}$)

Stage 3: Heat the liquid to boiling point

Liquid state: $C_{pB} = A + BT_B + CT_B^2 + DT_B^{-2}$ Equation 5-19

When A=30.55, B=0, C=0, D=0 and the maximum boiling temperature is 2400 K. The specific heat (c_{pB}) of Ag at the liquid state is 282.87 J Kg^{-1} K^{-1}. The enthalpy change at the melting point can be calculated from:

$$\Delta h = c_{pB} \cdot (T_B - T_M)$$ Equation 5-20

Stage 4: At the boiling point (at 2423 $\Leftarrow$ 20 K, the enthalpy of vaporisation is 2362.037×10^3 J Kg^{-1})

Stage 5: Heat the gas

Under the conditions of interested a typical arc temperature would be in the region of 10,000-20,000 K. However, the bulk gas flowing through the arc chamber has a much lower temperature of 3000-4000 K [11].

For ceramic arc chamber materials, the vapour layer may be at boiling temperature, but for Polymer it is definitely hotter. The data of Niemeyer [72] were consistent with polymer vapour temperature exceeding 3000 K.

Also, the experimental data of Ruchti and Niemeyer [73] in ablation mechanisms, the photoablation dominates causing a vapour temperature at 3400 +/- 200K. At this temperature ionization levels are relatively low and their contribution to the enthalpy can be neglected. The enthalpy supplied to heat a perfect gas is given by:

$$\Delta h = c_V \, \Delta T + PV \qquad\qquad \text{Equation 5-21}$$

$$\Delta h = c_v \, \Delta T + R \, \Delta T \qquad\qquad \text{Equation 5-22}$$

Where R of Ag is equal Ro/mass per mole of Ag (0.18 Kg), therefore R = 76.98 J Kg^{-1} K^{-1}, $\gamma = 5.98 \times 10^{-3}$, $c_v = c_p / \gamma = 47.29$ KJ Kg^{-1} K^{-1}.

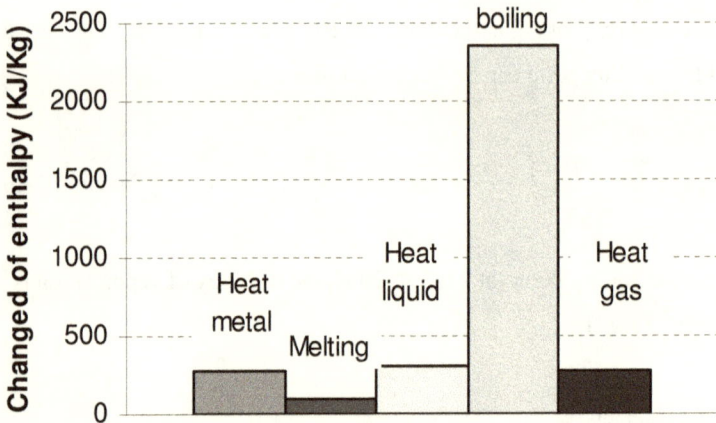

Figure 5.13: Changed of enthalpy from heat the metal until ionisation

Figure 5.13 shows the changed of enthalpy in each states from heat the metal until ionization. The changed of enthalpy at the boiling point is the maximum. At the melting the metal point the changed of enthalpy is the lowest.

The changed of enthalpy for heat liquid and gas is not much different. The total change of enthalpy from the summary of the heat metal, melting, heat liquid, boiling and heat gas is about 3450.46 kJ/kg.

5.5.1 Mass flow and volume flow

The arc power is defined as:

$$P_{arc}(t) = I(t) \cdot V(t)$$

Equation 5-23

From equation 5-15, the work rate is very little compared to the arc power. Thus, it is assumed that most of the arc power is used to melt the contact material and produce Ag gas. The arc power can be defended in term of enthalpy as follow:

$$P_{arc}(t) = \frac{dh}{dt} = \{ \int_{T_S}^{T_m} c_{ps}dt + dh(\text{melting}) + \int_{T_m}^{T_B} c_{pB}dt + dh(\text{boiling}) + \int_{T_B}^{T_G} c_v dt \} \cdot \frac{dm}{dt}$$

Equation 5-24

Where $\frac{dm}{dt}$ is the mass flow rate from the Ag contact material.

The density of the Ag gas is defined as:

$$\frac{P}{\rho_{Ag}} = R_{Ag} \cdot T$$

Equation 5-25

This gives the volume flow rate as:

$$\frac{dV}{dt} = \frac{1}{\rho_{Ag}} \cdot \frac{dm}{dt}$$

Equation 5-26

The density of Ag gas is approximately 0.539 kg m^{-3}. This can be converted to a flow velocity by assuming a channel cross sectional area A:

$$v_{fluid} = \frac{1}{A} \cdot \frac{dV}{dt}$$

Equation 5-27

The profile of the total arc power and the mass flow rate of Ag, at the period that the arc root moves off from the contact region are shown in Figure 5.14. When the channel cross section area in the arc chamber is about 150×10^{-6} mm^2. The volume flow rate and fluid velocity are shown in Figure 5.15.

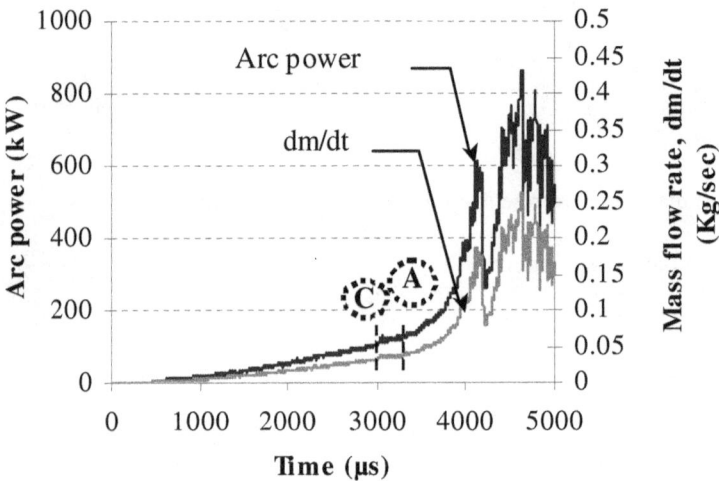

Figure 5.14: Arc power and mass flow rate (dm/dt) at contact opening velocity of 1 m/s, "C" is the point that the cathode root moves from the fixed contact, "A" is the point that the anode root moves from the moving contact, Ag/C step, ceramic arc chamber

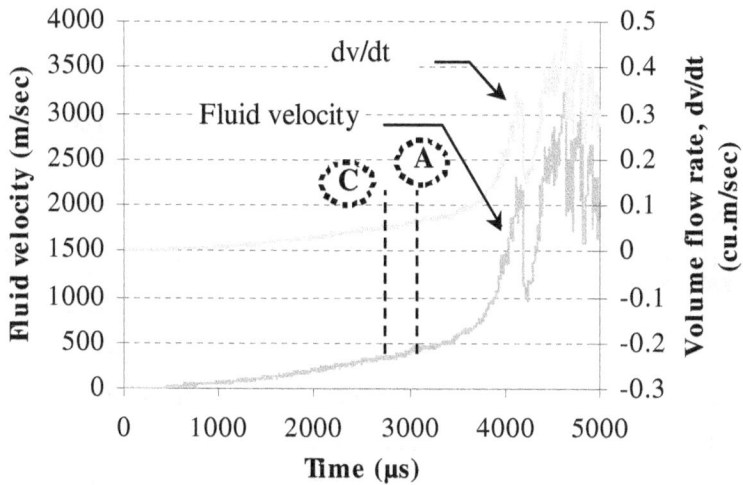

Figure 5.15: Volume flow rate and fluid velocity at contact opening velocity of 1 m/s, "C" is the point that the cathode root moves from the fixed contact, "A" is the point that the anode root moves from the moving contact, Ag/C step, ceramic arc chamber

Figure 5.14 show the arc power and mass flow rate at the point that the arc root moves from contact region. The arc power and mass flow rate when the anode root moves from the contact region is not much different from the cathode root.

The volume flow rate and fluid velocity with contact opening velocity 1 m/s are shown in Figure 5.15. The fluid velocity, at the point that the arc root moves from contact region, is about 480 m/s for cathode root and 500 m/s for anode root. The volume flow rate is about 0.05 m³/s.

5.6 Summary

Modeling and experimental data on the magnetic forces, gas dynamic forces and thermal energy influences in the contact region are used to improve the understanding of the physical principles and to identify the critical parameters governing the arc root motion in the contact region.

It is shown that there is a little significant difference in computed magnetic forces between Ag/C step and Ag/C flat contact materials. Additionally, the contact gap has a minimal effect on the magnetic forces on the arc root.

Both magnetic and gas dynamic models show higher forces at the point at which the anode root moves from the moving contact than when the cathode commutes from the fixed contact.

At the point that the arc root moves from the contact region, the arc voltage decreased, the arc current decreases, as the contact opening velocity is increased. The arc power at the point at which the arc root moves from the contact region on the different contact opening velocity are in the same range.

Energy considerations permit the calculation of the mass flow rate, volume flow rate, and fluid velocity in the arc chamber.

The results from this chapter are discussed in the chapter 7.

CHAPTER 6

ENERGY & TEMPERATURE RISE

6.1 Introduction

An emission spectrum can be obtained by heating an element. This gives the electron energy. The energies of these photons can be calculated using the following formulae:

$$E = h\,f \hspace{5cm} \text{Equation 6-1}$$

Where E is energy, h is Planck's constant (6.63 x 10-34 J s), f is frequency

$$C = \lambda\,f \hspace{5cm} \text{Equation 6-2}$$

Where C is the speed of light 3×10^8 m/s, λ is the wavelength

$$E = m \times cp \times \Delta T \hspace{4cm} \text{Equation 6-3}$$

Where m is mass, cp is the specific heat, ΔT temperature rise.

The contact area is approximate 0.3 cm width, 0.5 cm length and 0.6 cm dept. The volume of the contact region is approximate 0.09 cm^3.

The chemical elements characteristic density, specific heat and boiling point of Ag, N, O, C, H, Cu. Ti and Al are shown in Table 6.1.

Elements	Density at 293 K (g/cm^3)	Specific Heat, C (KJ/g K)	Boiling Point (K)
Ag	10.5	0.23	2485.15
N$_2$	1.165x10^{-3}	0.743	77.35

O_2	1.331×10^{-3}	0.659	90.15
C	2.62	0.46	5100.15
H_2	0.08988×10^{-3}	10.16	20.280005
Cu	8.96	0.39	2840.15
Ti	4.54	0.54	3560.15
Al	2.702	0.91	2740.15

Table 6.1: Chemical elements Characteristic

6.2 Polycarbonate Arc Chamber

Table 6.2 shows the 20 chemical elements in wavelength, frequency and electron energy in the polycarbonate arc chamber.

	Elements	Wavelength (nm)	Frequency	Electron energy (J)
1	N II	428	7.01E-04	4.65E-37
2	O II	428	7.01E-04	4.65E-37
3	C V	515	5.83E-04	3.86E-37
4	O IV	515	5.83E-04	3.86E-37
5	N V	515	5.83E-04	3.86E-37
6	O II	515	5.83E-04	3.86E-37
7	Ag I	521	5.76E-04	3.82E-37
8	Ag I	546	5.49E-04	3.64E-37
9	C III	567	5.29E-04	3.51E-37
10	O V	567	5.29E-04	3.51E-37
11	N II	656	4.57E-04	3.03E-37
12	H	656	4.57E-04	3.03E-37
13	O II	656	4.57E-04	3.03E-37
14	O I	776	3.87E-04	2.56E-37
15	N II	776	3.87E-04	2.56E-37

16	C II	766	3.92E-04	2.60E-37
17	Cu II	766	3.92E-04	2.60E-37
18	N I	821	3.65E-04	2.42E-37
19	O I	821	3.65E-04	2.42E-37
20	O II	821	3.65E-04	2.42E-37

Table 6.2: Electron Energy, Frequency and Spectrum wavelength in the Polycarbonate arc chamber

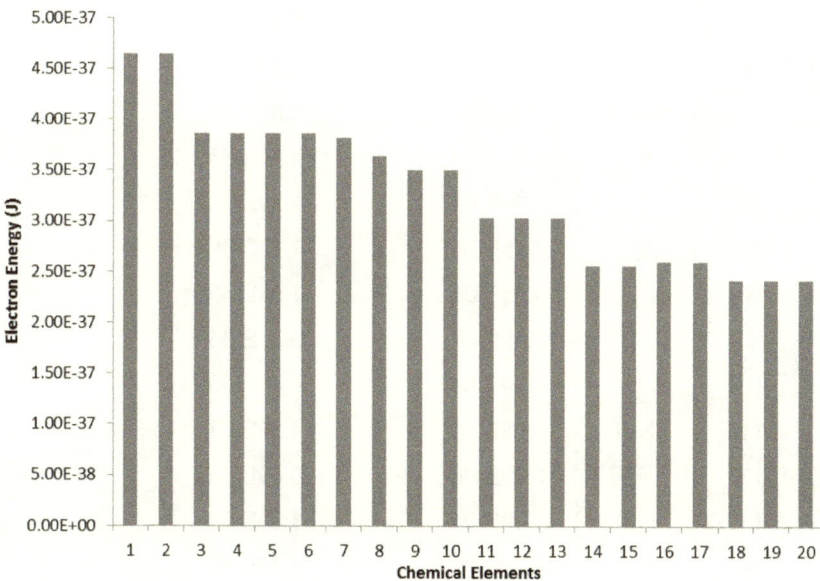

Figure 6.1: Electron energy in the Polycarbonate arc chamber

Figure 6.1 shows the relationship between electron energy and chemical elements in the Polycarbonate arc chamber. There are 20 chemical elements in the Polycarbonate arc chamber. The electron energy for each chemical elements decrease as the wavelength increases.

The maximum energy is 4.65×10^{-37} J from the chemical element N II and O II at the wavelength 428 nm. The minimum energy is 2.42×10^{-37} J from the chemical element

N I, O I and O II at the wavelength 821 nm. The average energy in the Polycarbonate arc chamber is 3.30×10^{-37} J.

Table 6.3 shows the Mass and Temperature rise of the 20 chemical elements in the Polycarbonate arc chamber.

	Elements	Mass (g)	Temperature rise, ΔT (K)
1	N II	1.05E-04	5.97E-33
2	O II	1.20E-04	5.89E-33
3	C V	2.36E-01	3.56E-36
4	O IV	1.20E-04	4.89E-33
5	N V	1.05E-04	4.96E-33
6	O II	1.20E-04	4.89E-33
7	Ag I	9.45E-01	1.76E-36
8	Ag I	9.45E-01	1.68E-36
9	C III	2.36E-01	3.23E-36
10	O V	1.20E-04	4.44E-33
11	N II	1.05E-04	3.89E-33
12	H	8.09E-06	3.69E-33
13	O II	1.20E-04	3.84E-33
14	O I	1.20E-04	3.25E-33
15	N II	1.05E-04	3.29E-33
16	C II	2.36E-01	2.39E-36
17	Cu II	8.06E-01	8.26E-37
18	N I	1.05E-04	3.11E-33
19	O I	1.20E-04	3.07E-33
20	O II	1.20E-04	3.07E-33

Table 6.3: Mass and Temperature rise in Polycarbonate arc chamber

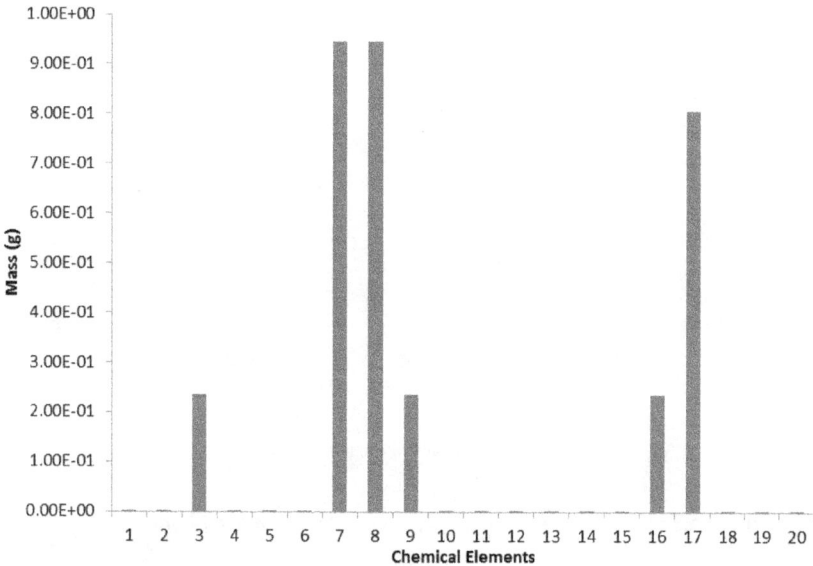

Figure 6.2: Chemical element Mass in the Polycarbonate arc chamber

Figure 6.2 shows the relationship between Mass and Chemical elements in the Polycarbonate arc chamber. There are 20 chemical elements in the Polycarbonate arc chamber. The maximum Mass in the Polycarbonate arc chamber belongs to the chemical element Ag I.

The Mass of the chemical element Cu II is a little bit lower than the chemical element Ag I. The chemical element H is the minimum mass in the Polycarbonate arc chamber. The Mass average in the Polycarbonate arc chamber is 0.17 g.

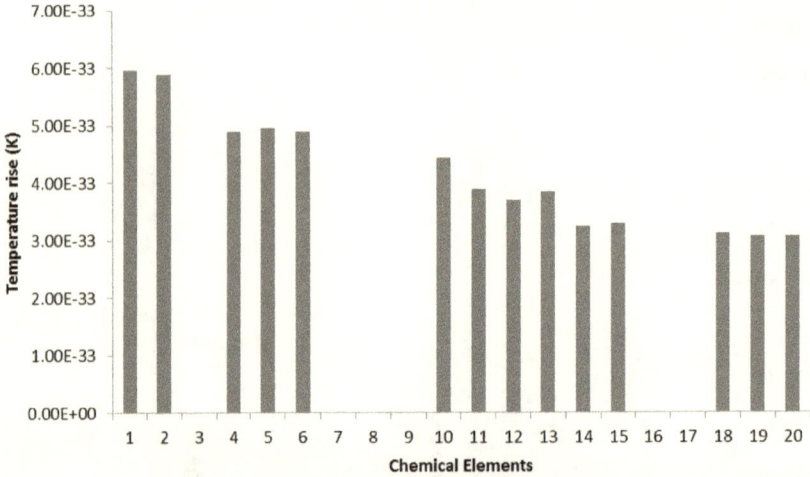

Figure 6.3: Temperature rise in the Polycarbonate arc chamber

Figure 6.3 shows the relationship between Temperature rise and Chemical elements in the Polycarbonate arc chamber. There are 20 chemical elements in the Polycarbonate arc chamber. The maximum temperature rise in the Polycarbonate arc chamber is the chemical element N II.

The minimum temperature rise is the chemical element Cu II. The temperature rise for the Ag I at the wavelength 521 nm and 546 nm, this is a little bit more than the minimum temperature rise. However, the Ag I energy is much higher than the chemical element Cu II. The average temperature rise in the Polycarbonate arc chamber is 2.91×10^{-33} K.

6.3 Ceramic Arc Chamber

Table 6.4 shows the 26 chemical elements in wavelength, frequency and electron energy in the ceramic arc chamber.

	Elements	Wavelength (nm)	Frequency	Electron energy
1	O II	500	6.00E-04	3.98E-37
2	O IV	500	6.00E-04	3.98E-37
3	Ti III	500	6.00E-04	3.98E-37
4	Al IV	500	6.00E-04	3.98E-37
5	N III	500	6.00E-04	3.98E-37
6	Ag I	521	5.76E-04	3.82E-37
7	Ag I	546	5.49E-04	3.64E-37
8	C III	567	5.29E-04	3.51E-37
9	O V	567	5.29E-04	3.51E-37
10	O IV	594	5.05E-04	3.35E-37
11	C II	594	5.05E-04	3.35E-37
12	Al IV	594	5.05E-04	3.35E-37
13	N IV	594	5.05E-04	3.35E-37
14	N II	656	4.57E-04	3.03E-37
15	H	656	4.57E-04	3.03E-37
16	O II	656	4.57E-04	3.03E-37
17	N I	746	4.02E-04	2.67E-37
18	N II	746	4.02E-04	2.67E-37
19	O III	746	4.02E-04	2.67E-37
20	O I	776	3.87E-04	2.56E-37
21	N II	776	3.87E-04	2.56E-37
22	C II	776	3.87E-04	2.56E-37
23	Cu II	776	3.87E-04	2.56E-37
24	N I	821	3.65E-04	2.42E-37
25	O I	821	3.65E-04	2.42E-37
26	O II	821	3.65E-04	2.42E-37

Table 6.4: Electron Energy, Frequency and Spectrum wavelength in ceramic arc chamber

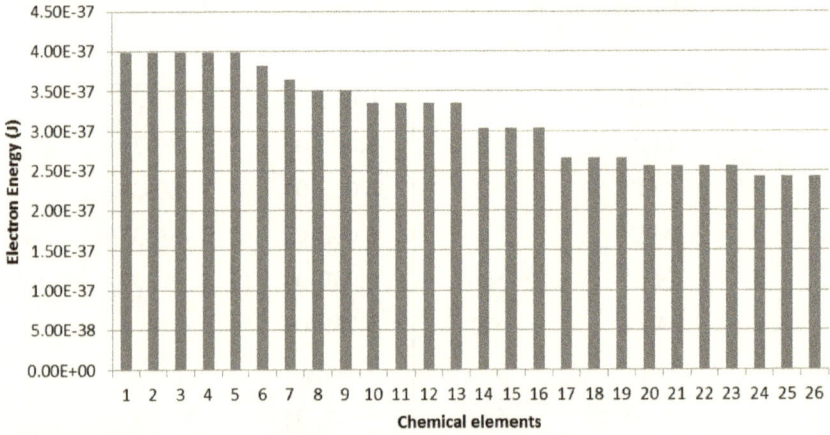

Figure 6.4: Electron energy in ceramic arc chamber

Figure 6.4 shows the relationship between electron energy and chemical elements in the ceramic arc chamber. There are 26 chemical elements in the ceramic arc chamber. The electron energy for each chemical elements decrease as the wavelength increases.

The maximum energy is 3.98×10^{-37} J from the chemical element O II, O IV, Ti III, Al IV and N III at the wavelength 500 nm. The minimum energy is 2.42×10^{-37} J from the chemical element N I, O I and O II at the wavelength 821 nm. The average energy in the ceramic arc chamber is 3.17×10^{-37} J.

Table 6.5 shows the Mass and Temperature rise of the 20 chemical elements in the Ceramic arc chamber.

	Elements	Mass (g)	Temperature rise, ΔT (K)
1	O II	1.20E-04	5,04E-33
2	O IV	1.20E-04	5.04E-33
3	Ti III	4.09E-01	1.80E-36
4	Al IV	2.43E-01	1.80E-36
5	N III	1.05E-04	5.11E-33

6	Ag I	9.45E-01	1.76E-36
7	Ag I	9.45E-01	1.68E-36
8	C III	2.36E-01	3.23E-36
9	O V	1.20E-04	4.44E-33
10	O IV	1.20E-04	4.24E-33
11	C II	2.36E-01	3.09E-36
12	Al IV	2.43E-01	1.51E-36
13	N IV	1.05E-04	4.30E-33
14	N II	1.05E-04	3.89E-33
15	H	8.09E-06	3.69E-33
16	O II	1.20E-04	3.84E-33
17	N I	1.05E-04	3.42E-33
18	N II	1.05E-04	3.42E-33
19	O III	1.20E-04	3.38E-33
20	O I	1.20E-04	3.25E-33
21	N II	1.05E-04	3.29E-33
22	C II	2.36E-01	2.36E-36
23	Cu II	8.06E-01	8.15E-37
24	N I	1.05E-04	3.11E-33
25	O I	1.20E-04	3.07E-33
26	O II	1.20E-04	3.07E-33

Table 6.5: Mass and Temperature rise in ceramic arc chamber

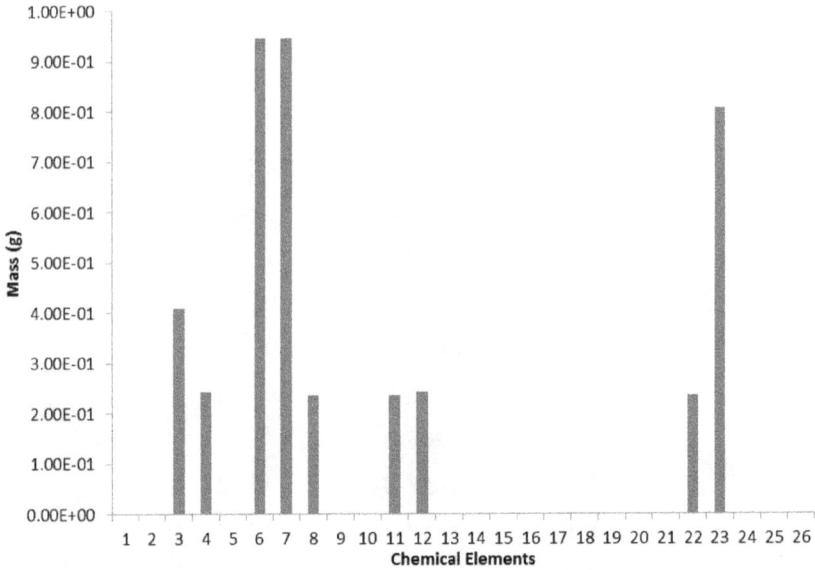

Figure 6.5: Chemical element Mass in the ceramic arc chamber

Figure 6.5 shows the relationship between mass and chemical elements in the ceramic arc chamber. There are 26 chemical elements in the ceramic arc chamber. The Mass of Ag I is dominated in the ceramic arc chamber which is the maximum Mass in the ceramic arc chamber while H is the smallest Mass. The Mass average in the ceramic arc chamber is 0.165 g.

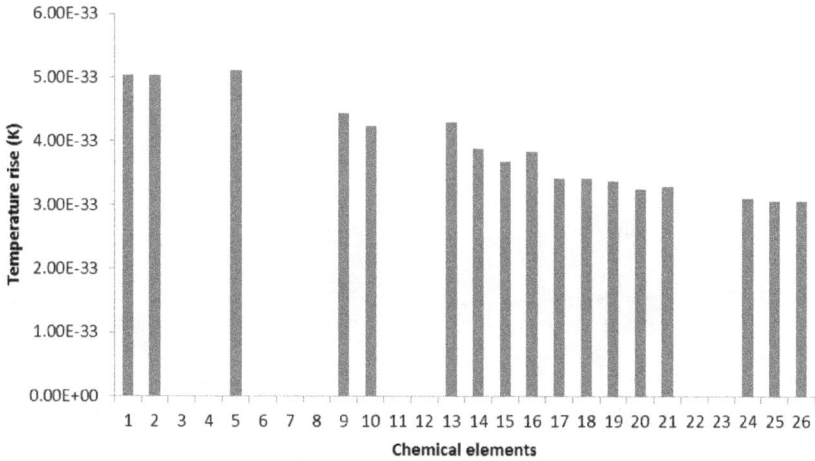

Figure 6.6: Temperature rise in ceramic arc chamber

Figure 6.6 shows the relationship between Temperature rise and chemical elements in the ceramic arc chamber. There are 26 chemical elements in the ceramic arc chamber. The maximum temperature rise in the ceramic arc chamber is from the chemical element N III. The minimum temperature rise is from the chemical element Cu II.

6.4 Contact velocity 1 m/s

Table 6.6 shows the 19 chemical elements in wavelength, frequency and electron energy when the contact opening velocity 1 m/s

	Elements	Wavelength (nm)	Frequency	Electron energy (J)
1	C V	515	5.83E-04	3.86E-37
2	O II	515	5.83E-04	3.86E-37
3	O IV	515	5.83E-04	3.86E-37
4	N V	515	5.83E-04	3.86E-37
5	Ag I	521	5.76E-04	3.82E-37

6	Ag I	546	5.49E-04	3.64E-37
7	N II	656	4.57E-04	3.03E-37
8	H	656	4.57E-04	3.03E-37
9	O II	656	4.57E-04	3.03E-37
10	O I	776	3.87E-04	2.56E-37
11	N II	776	3.87E-04	2.56E-37
12	C II	776	3.87E-04	2.56E-37
13	Cu II	776	3.87E-04	2.56E-37
14	N I	821	3.65E-04	2.42E-37
15	O I	821	3.65E-04	2.42E-37
16	O II	821	3.65E-04	2.42E-37
17	N I	824	3.64E-04	2.41E-37
18	O I	824	3.64E-04	2.41E-37
19	O II	824	3.64E-04	2.41E-37

Table 6.6: Electron Energy, Frequency and Spectrum wavelength when the contact opening velocity 1 m/s

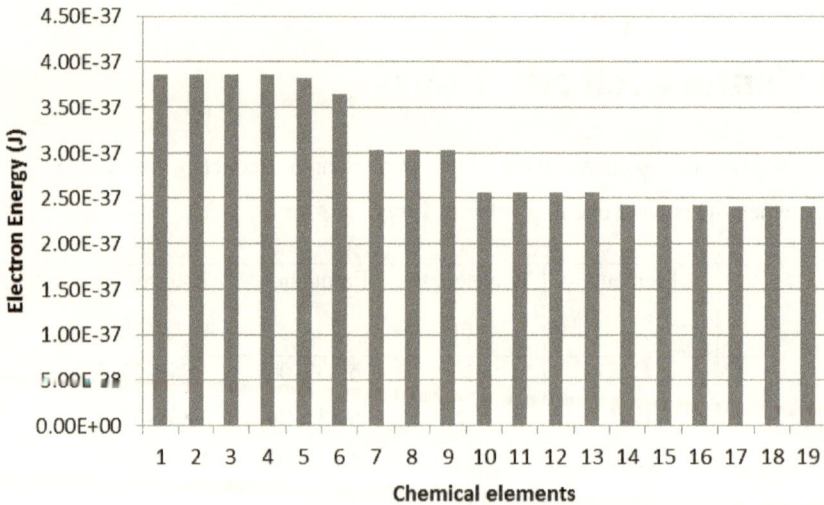

Figure 6.7: Electron energy when the contact opening velocity 1 m/s

Figure 6.7 shows the relationship between electron energy and chemical elements when the contact opening velocity 1 m/s. There are 19 chemical elements when the contact opening velocity 1 m/s. The energy decreases as the wavelength increases but not significantly.

The maximum energy is 3.86×10^{-37} J at the wavelength 515 nm. At the wavelength 824 nm is the lowest energy, which it is 2.41×10^{-37} J from the chemical element N I, O I and O II. The average energy for the contact opening velocity 1m/s is 2.99×10^{-37} J.

Table 6.7 shows the Mass and Temperature rise of the 19 chemical elements when the contact opening velocity 1 m/s.

	Elements	Mass (g)	Temperature rise, ΔT (K)
1	C V	2.36E-01	3.56E-36
2	O II	1.20E-04	4.89E-33
3	O IV	1.20E-04	4.89E-33
4	N V	1.05E-04	4.96E-33
5	Ag I	9.45E-01	1.76E-36
6	Ag I	9.45E-01	1.68E-36
7	N II	1.05E-04	3.89E-33
8	H	8.09E-06	3.69E-33
9	O II	1.20E-04	3.84E-33
10	O I	1.20E-04	3.25E-33
11	N II	1.05E-04	3.29E-33
12	C II	2.36E-01	2.36E-36
13	Cu II	8.06E-01	8.15E-37
14	N I	1.05E-04	3.11E-33
15	O I	1.20E-04	3.07E-33
16	O II	1.20E-04	3.07E-33
17	N I	1.05E-04	3.10E-33
18	O I	1.20E-04	3.06E-33

19	O II	1.20E-04	3.06E-33

Table 6.7: Mass and Temperature rise wavelength when the contact opening velocity 1 m/s

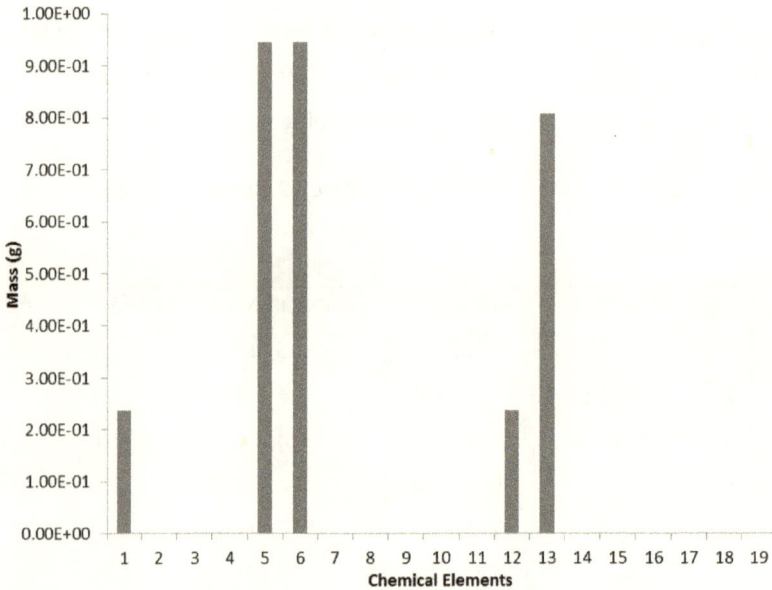

Figure 6.8: Chemical element Mass when the contact opening velocity 1 m/s

Figure 6.8 shows the relationship between Mass and chemical elements when the contact opening velocity 1 m/s. There are 19 chemical elements when the contact opening velocity 1 m/s. The chemical elements of Ag I and Cu II at the wavelength 521, 546, 776 nm are outstanding showing the domination of the mass when contact opening velocity at 1 m/s.

The maximum Mass is the chemical element from the Ag I and the second high Mass is the chemical element from the Cu II. The minimum Mass is the chemical element H and the average Mass when the contact opening velocity 1 m/s is 0.167 g.

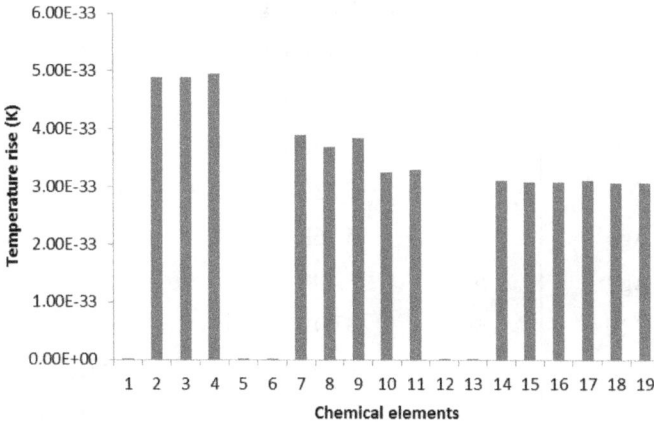

Figure 6.9: Temperature rise when the contact opening velocity 1 m/s

Figure 6.9 shows the relationship between Temperature rise and chemical elements when the contact opening velocity 1 m/s. There are 19 chemical elements when the contact opening velocity 1 m/s. The temperature rise is between 3×10^{-33} K and 5×10^{-33} K for all chemical elements.

The highest temperature rise is the chemical element N V. This is not very much higher than the chemical elements O II and O IV. The average temperature rise for the contact opening velocity at 1 m/s is approximate 2.69×10^{-33} K. The minimum temperature rise is at the wavelength 776 nm of the Cu II chemical element.

6.5 Contact velocity 10 m/s

Table 6.8 shows the 20 chemical elements in wavelength, frequency and electron energy when the contact opening velocity 10 m/s.

	Elements	Wavelength (nm)	Frequency	Electron energy (J)
1	N II	428	7.01E-04	4.65E-37

2	O II	428	7.01E-04	4.65E-37
3	C V	515	5.83E-04	3.86E-37
4	O IV	515	5.83E-04	3.86E-37
5	N V	515	5.83E-04	3.86E-37
6	O II	515	5.83E-04	3.86E-37
7	Ag I	521	5.76E-04	3.82E-37
8	Ag I	546	5.49E-04	3.64E-37
9	C III	567	5.29E-04	3.51E-37
10	O V	567	5.29E-04	3.51E-37
11	N II	656	4.57E-04	3.03E-37
12	H	656	4.57E-04	3.03E-37
13	O II	656	4.57E-04	3.03E-37
14	O I	776	3.87E-04	2.56E-37
15	N II	776	3.87E-04	2.56E-37
16	C II	766	3.92E-04	2.60E-37
17	Cu II	766	3.92E-04	2.60E-37
18	N I	821	3.65E-04	2.42E-37
19	O I	821	3.65E-04	2.42E-37
20	O II	821	3.65E-04	2.42E-37

Table 6.8: Electron Energy, Frequency and Spectrum wavelength when the contact opening velocity 10 m/s

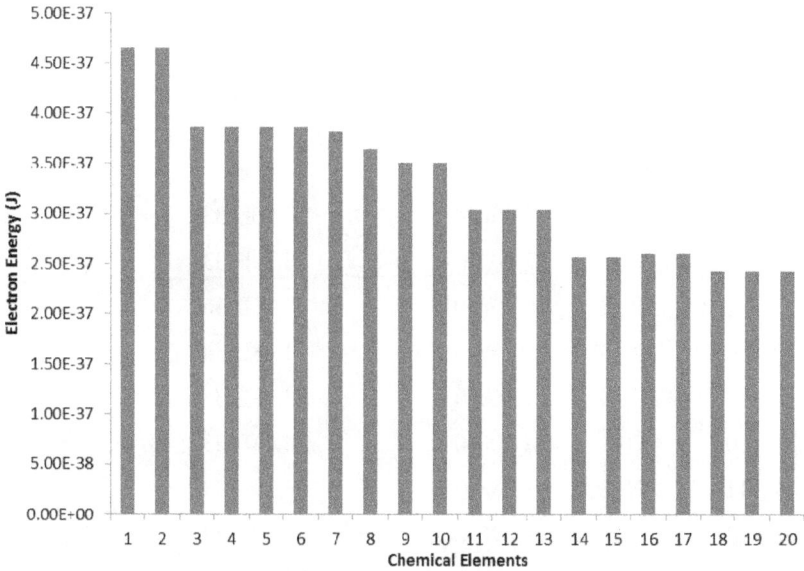

Figure 6.10: Electron energy when the contact opening velocity 10 m/s

Figure 6.10 shows the relationship between electron energy and chemical when the contact opening velocity 10 m/s. There are 20 chemical elements when the contact opening velocity 10 m/s. The electron energy for each chemical elements decrease as the wavelength increases. However, the gap between the lowest and highest energy is about $2.50 \times 10^{-37} - 4.50 \times 10^{-37}$ J.

The maximum energy is 4.65×10^{-37} J from the chemical element N II and O II at the wavelength 428 nm. The minimum energy is 2.42×10^{-37} J from the chemical element N I, O I and O II at the wavelength 821 nm. The average energy when the contact opening velocity 10 m/s is 3.30×10^{-37} J.

Table 6.9 shows the Mass and Temperature rise of the 20 chemical elements when the contact opening velocity 10 m/s

	Elements	Mass (g)	Temperature rise, ΔT (K)
1	N II	1.05E-04	5.97E-33
2	O II	1.20E-04	5.89E-33
3	C V	2.36E-01	3.56E-36
4	O IV	1.20E-04	4.89E-33
5	N V	1.05E-04	4.96E-33
6	O II	1.20E-04	4.89E-33
7	Ag I	9.45E-01	1.76E-36
8	Ag I	9.45E-01	1.68E-36
9	C III	2.36E-01	3.23E-36
10	O V	1.20E-04	4.44E-33
11	N II	1.05E-04	3.89E-33
12	H	8.09E-06	3.69E-33
13	O II	1.20E-04	3.84E-33
14	O I	1.20E-04	3.25E-33
15	N II	1.05E-04	3.29E-33
16	C II	2.36E-01	2.39E-36
17	Cu II	8.06E-01	8.26E-37
18	N I	1.05E-04	3.11E-33
19	O I	1.20E-04	3.07E-33
20	O II	1.20E-04	3.07E-33

Table 6.9: Mass and Temperature rise when the contact opening velocity 10 m/s

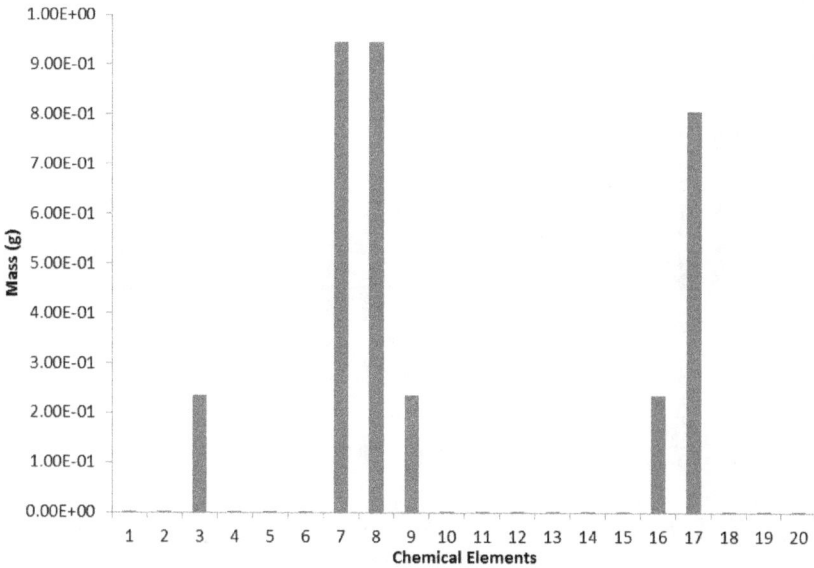

Figure 6.11: Chemical element Mass when the contact opening velocity 10 m/s

Figure 6.11 shows the relationship between Mass and chemical elements when the contact opening velocity 10 m/s. There are 20 chemical elements when the contact opening velocity 10 m/s. The maximum Mass when the contact opening velocity 10 m/s is the chemical element Ag I.

The Mass of Cu II is lower than Ag I about 6.7 g. The chemical element H is the minimum Mass when the contact opening velocity 10 m/s. The Mass average when the contact opening velocity 10 m/s, is 0.17 g.

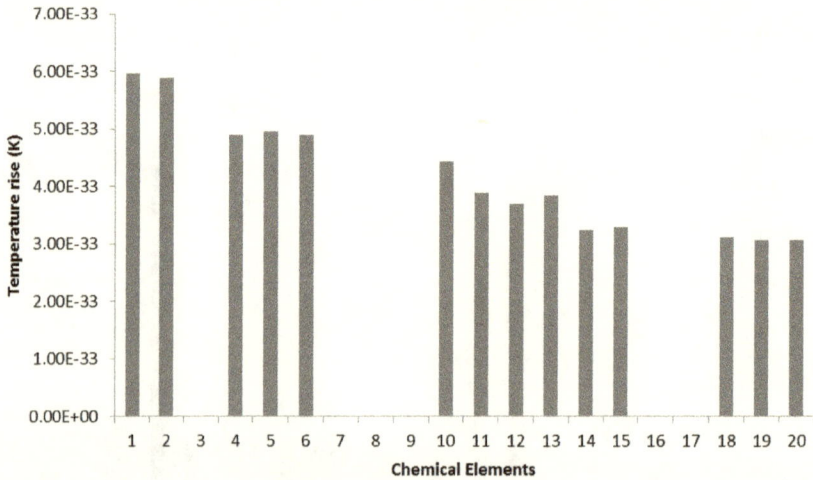

Figure 6.12: Temperature rise when the contact opening velocity 10 m/s

Figure 6.12 shows the relationship between Temperature rise and Chemical elements when the contact opening velocity 10 m/s. There are 20 chemical elements when the contact opening velocity 10 m/s. The maximum temperature rise when the contact opening velocity 10 m/s is the chemical element N II, but it is not significantly higher than the chemical element O II.

The minimum temperature rise is from the chemical element Cu II. The average temperature rise when the contact opening velocity 10 m/s, is 2.91×10^{-33} K.

6.6 Contact Material Ag/C step

Table 6.10 shows the 25 chemical elements in wavelength, frequency and electron energy when using Ag/C step contact material.

	Elements	Wavelength (nm)	Frequency	Electron energy (J)
1	O II	500	6.00E-04	3.98E-37

2	O IV	500	6.00E-04	3.98E-37
3	Ti VIII	500	6.00E-04	3.98E-37
4	Al IV	500	6.00E-04	3.98E-37
5	N III	500	6.00E-04	3.98E-37
6	Ag I	546	5.49E-04	3.64E-37
7	C III	567	5.29E-04	3.51E-37
8	O V	567	5.29E-04	3.51E-37
9	O IV	594	5.05E-04	3.35E-37
10	C II	594	5.05E-04	3.35E-37
11	Al IV	594	5.05E-04	3.35E-37
12	N IV	594	5.05E-04	3.35E-37
13	N II	656	4.57E-04	3.03E-37
14	H	656	4.57E-04	3.03E-37
15	O II	656	4.57E-04	3.03E-37
16	N I	746	4.02E-04	2.67E-37
17	N II	746	4.02E-04	2.67E-37
18	O III	746	4.02E-04	2.67E-37
19	O I	776	3.87E-04	2.56E-37
20	N II	776	3.87E-04	2.56E-37
21	C II	776	3.87E-04	2.56E-37
22	Cu II	776	3.87E-04	2.56E-37
23	N I	821	3.65E-04	2.42E-37
24	O I	821	3.65E-04	2.42E-37
25	O II	821	3.65E-04	2.42E-37

Table 6.10: Electron Energy, Frequency and Spectrum wavelength when using Ag/C step contact material

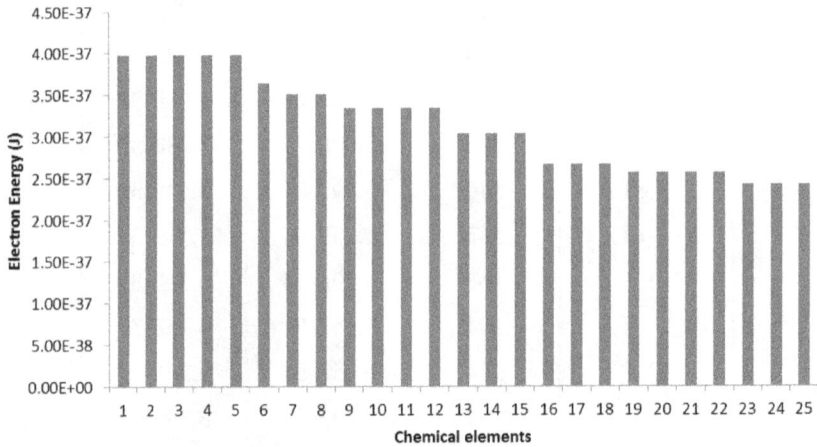

Figure 6.13: Electron energy when using Ag/C step contact material

Figure 6.13 shows the relationship between electron energy and chemical elements when using Ag/C step contact material. There are 25 chemical elements when using Ag/C step contact material. At the wavelength 500 nm, it shows the maximum electron energy at 3.98×10^{-37} J from the chemical elements O II, O, IV, Ti VIII, Al IV and N III.

The average energy for all chemical elements when using the Ag/C step contact material is about 3.14×10^{-37} J. The lowest energy is at the wavelength 821 nm for chemical elements N I, O I and O II about 2.42×10^{-37} J.

Table 6.11 shows the Mass and Temperature rise of the 25 chemical elements when using Ag/C step contact material.

	Elements	Mass (g)	Temperature rise, ΔT (K)
1	O II	1.20E-04	5.04E-33
2	O IV	1.20E-04	5.04E-33
3	Ti VIII	4.09E-01	1.80E-36
4	Al IV	2.43E-01	1.80E-36

5	N III	1.05E-04	5.11E-33
6	Ag I	9.45E-01	1.68E-36
7	C III	2.36E-01	3.23E-36
8	O V	1.20E-04	4.44E-33
9	O IV	1.20E-04	4.24E-33
10	C II	2.36E-01	3.09E-36
11	Al IV	2.43E-01	1.51E-36
12	N IV	1.05E-04	4.30E-33
13	N II	1.05E-04	3.89E-33
14	H	8.99E-05	3.32E-34
15	O II	1.20E-04	3.84E-33
16	N I	1.05E-04	3.42E-33
17	N II	1.05E-04	3.42E-33
18	O III	1.20E-04	3.38E-33
19	O I	1.20E-04	3.25E-33
20	N II	1.05E-04	3.29E-33
21	C II	2.36E-01	2.36E-36
22	Cu II	8.06E-01	8.15E-37
23	N I	1.05E-04	3.11E-33
24	O I	1.20E-04	3.07E-33
25	O II	1.20E-04	3.07E-33

Table 6.11: Mass and Temperature rise wavelength when using Ag/C step contact material

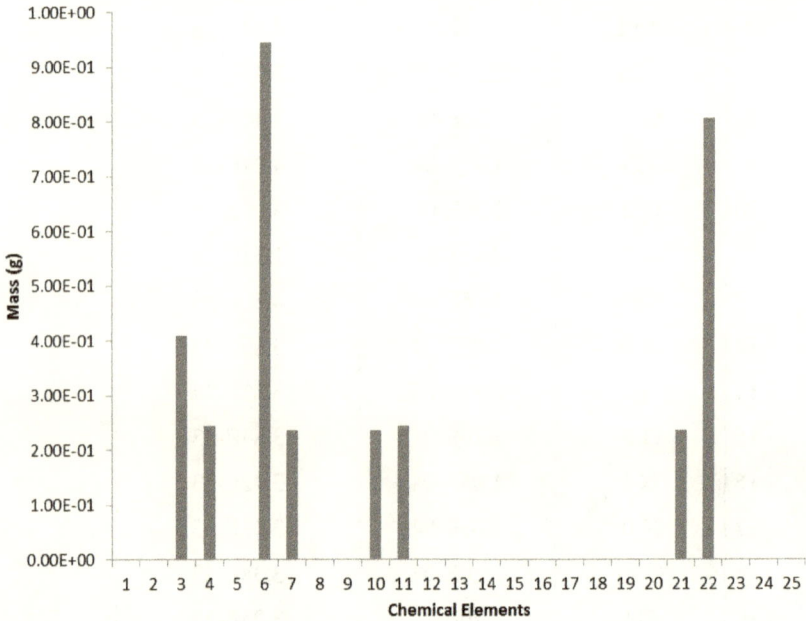

Figure 6.14: Chemical element and Mass when using Ag/C step contact material

Figure 6.14 shows the relationship between Mass and chemical elements when using Ag/C step contact material. There are 25 chemical elements when using Ag/C step contact material.

The Mass of the chemical elements Ag I and Cu II are higher than other chemical elements about 4 times. The chemical element H has the lowest Mass when using Ag/C step contact material. The average Mass of Ag/C contact material is 0.134g.

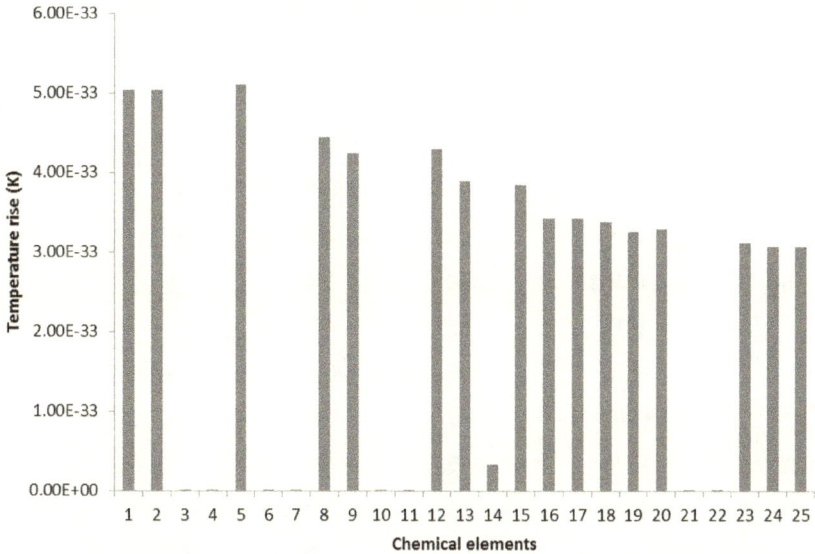

Figure 6.15: Temperature rise when using Ag/C step contact material

Figure 6.15 shows the relationship between Temperature rise and chemical elements when using Ag/C step contact material. There are 25 chemical elements when using Ag/C step contact material. The highest temperature rise belong to chemical element N III at the wavelength 500 nm. The temperature rise of the chemical element O II and O IV are a little bit lower than the chemical element NIII. The lowest temperature rise is the chemical element Cu II at the wavelength 776 nm. The average temperature rise for the Ag/C step contact material is 2.49×10^{-33} K.

6.7 Cu Punch Contact Material

Table 6.12 shows the 19 chemical elements in wavelength, frequency and electron energy when using the Cu Punch contact material.

	Elements	Wavelength (nm)	Frequency	Electron Energy (J)
1	N II	428	7.01E-04	4.65E-37
2	O II	428	7.01E-04	4.65E-37
3	O II	500	6.00E-04	3.98E-37
4	O IV	500	6.00E-04	3.98E-37
5	Ti VIII	500	6.00E-04	3.98E-37
6	Al IV	500	6.00E-04	3.98E-37
7	N III	500	6.00E-04	3.98E-37
8	C III	567	5.29E-04	3.51E-37
9	O V	567	5.29E-04	3.51E-37
10	N II	656	4.57E-04	3.03E-37
11	H	656	4.57E-04	3.03E-37
12	O II	656	4.57E-04	3.03E-37
13	O I	776	3.87E-04	2.56E-37
14	N II	776	3.87E-04	2.56E-37
15	C II	776	3.87E-04	2.56E-37
16	Cu II	776	3.87E-04	2.56E-37
17	N I	821	3.65E-04	2.42E-37
18	O I	821	3.65E-04	2.42E-37
19	O II	821	3.65E-04	2.42E-37

Table 6.12: Electron Energy, Frequency and Spectrum wavelength when using Cu Punch contact material

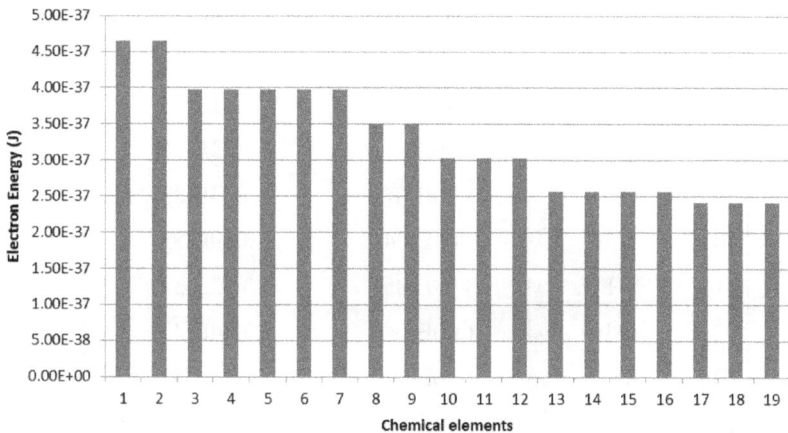

Figure 6.16: Electron energy when using Cu Punch contact material

Figure 6.16 shows the relationship between electron energy and chemical elements when using Cu Punch contact material. There are 19 chemical elements when using Cu Punch contact material. The energy decreases as the wavelength increase, but the energy range is approximate 2.45×10^{-37} - 4.60×10^{-37} J.

The highest energy is 4.65×10^{-37} J at the wavelength 428 nm of the chemical element N II and O II. The lowest energy is 2.42×10^{-37} J at the wavelength 821 nm of the chemical element N I, O I and N II. The average energy of the Cu punch contact material is 3.31×10^{-37} J.

Table 6.13 shows the Mass and Temperature rise of the 19 chemical elements when using Cu Punch contact material.

	Elements	Mass (g)	Temperature rise, ΔT (K)
1	N II	1.05E-04	5.97E-33
2	O II	1.20E-04	5.89E-33
3	O II	1.20E-04	5.04E-33
4	O IV	1.20E-04	5.04E-33

5	Ti VIII	4.09E-01	1.80E-36
6	Al IV	2.43E-01	1.80E-36
7	N III	1.05E-04	5.11E-33
8	C III	2.36E-01	3.23E-36
9	O V	1.20E-04	4.44E-33
10	N II	1.05E-04	3.89E-33
11	H	8.99E-05	3.32E-34
12	O II	1.20E-04	3.84E-33
13	O I	1.20E-04	3.25E-33
14	N II	1.13E-01	3.06E-36
15	C II	2.36E-01	2.36E-36
16	Cu II	8.06E-01	8.15E-37
17	N I	1.05E-04	3.11E-33
18	O I	1.20E-04	3.07E-33
19	O II	1.20E-04	3.07E-33

Table 6.13: Mass and Temperature rise when using Cu Punch contact material

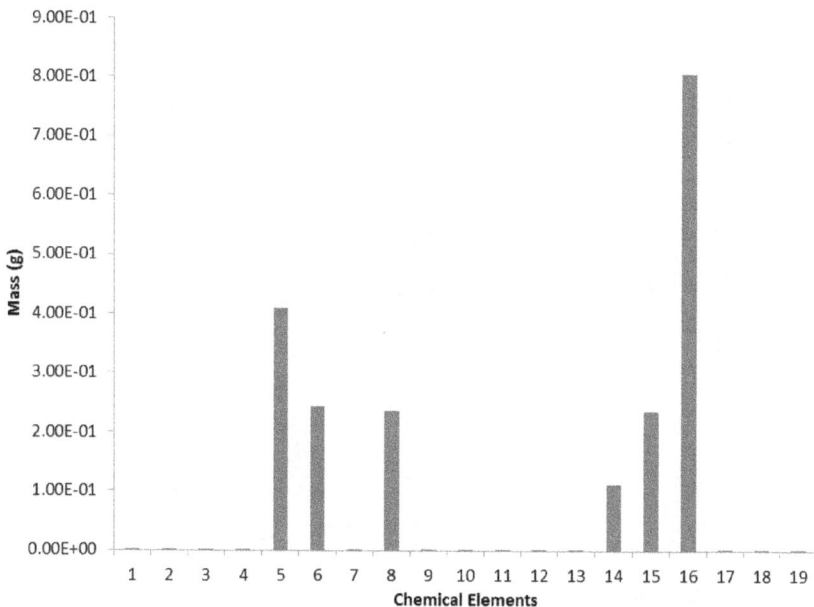

Figure 6.17: Chemical element mass when using Cu Punch contact material

Figure 6.17 shows the relationship between Mass and chemical elements when using Cu Punch contact material. There are 19 chemical elements when using Cu Punch contact material. The maximum Mass is the chemical element Cu II at the wavelength 776 nm. The minimum Mass is the chemical element H at the wavelength 656 nm. The average Mass of the Cu punch contact material is 0.108 g.

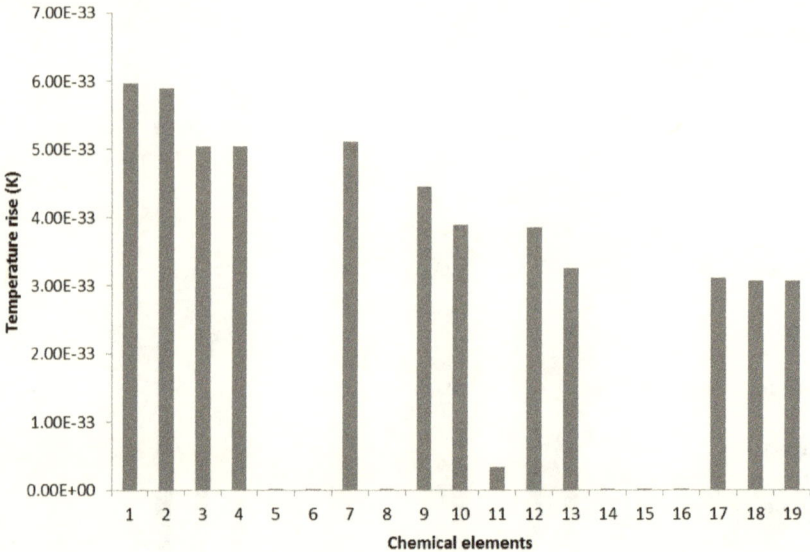

Figure 6.18: Temperature rise when using Cu Punch contact material

Figure 6.18 shows the relationship between Temperature rise and chemical elements when using Cu Punch contact material. There are 19 chemical elements when using Cu Punch contact material. The maximum temperature rise for the Cu punch contact material is the chemical elements N II at the wavelength 428 nm but the second highest is the chemical element O II, which it is nearly the same temperature rise as the chemical element N II. The minimum temperature rise is the chemical element Cu II at the wavelength 776 nm. The average temperature rise is 2.74×10^{-33} K.

6.8 Summary

6.8.1 Polycarbonate and Ceramic arc chamber

Both the polycarbonate arc chamber and the ceramic arc chamber show that the electron energy is decreased as the wavelength increases. The polycarbonate arc chamber has 20 chemical elements, while the ceramic arc chamber has 26 chemical elements. The maximum energy polycarbonate arc chamber is higher than the ceramic arc chamber about 0.67×10^{-37} J.

Both the polycarbonate arc chamber and ceramic arc chamber have the same minimum energy at 2.42×10^{-37} J and have the same chemical elements N I, O I and O II at the wavelength 821 nm. The wavelength at the maximum energy of the polycarbonate arc chamber is lower than the ceramic arc chamber about 72 nm. The polycarbonate arc chamber and ceramic arc chamber have the same chemical elements at the maximum energy which it is O II. The average energy in the Polycarbonate arc chamber is higher than the ceramic arc chamber 0.13×10^{-37} J

The maximum Mass in the Polycarbonate arc chamber is the same as the ceramic arc chamber. This comes from the chemical element Ag I which it is the main ingredient of the contact material. The chemical element H is the minimum Mass in both the Polycarbonate and the ceramic arc chamber. The Mass average in the Polycarbonate arc chamber is approximately 0.17 g. The number of the chemical elements in the ceramic arc chamber is more than the polycarbonate arc chamber about 4 chemical elements, but the Mass average in the ceramic arc chamber is 0.005 g less than the Polycarbonate arc chamber.

The maximum temperature rise in the polycarbonate arc chamber is higher than the ceramic arc chamber. From the same chemical element Cu II, the minimum temperature rise of the polycarbonate arc chamber more than the ceramic arc chamber. The average temperature rise in the polycarbonate arc chamber is higher than the ceramic arc chamber.

In the polycarbonate arc chamber, the temperature rise for the chemical element Ag I at the wavelength 521 nm and 546 nm is a little bit more than the minimum temperature rise. However, the Ag I electron energy is much higher than the chemical element Cu II.

6.8.2 Contact opening velocity at 10 and 1 m/s

There are 20 chemical elements when the contact opening velocity 10 m/s and 19 chemical elements when the contact opening velocity 1 m/s. The electron energy for each chemical elements decrease as the wavelength increases. The maximum energy when contact opening velocity at 10 m/s, is about 0.79×10^{-37} J higher than that contact opening velocity 1 m/s. Both have the same minimum energy at 2.42×10^{-37} J of the chemical element N I, O I and O II at the wavelength 821-824 nm. The average energy when the contact opening velocity 10 m/s is higher than the contact opening velocity 1m/s about 0.31×10^{-37} J.

The maximum Mass is the chemical element from the Ag I and the second high Mass is the chemical element from the Cu II. The minimum Mass is the chemical element H. The average Mass when the contact opening velocity 1 m/s is 0.167 g and 0.17g for the contact opening velocity 10 m/s. The maximum Mass, when the contact opening velocity 10 and 1 m/s is the chemical element Ag I. The chemical elements of Ag I and Cu II at the wavelength 521, 546, 776 nm are outstanding show the domination of the Mass when contact opening velocity at 1 m/s.

When the contact opening velocity 1 m/s, the temperature rise for all chemical elements are between 3x10-33 K and 5x10-33 K. The maximum temperature rise when the contact opening velocity 10 m/s is the chemical element N II, but it is not significantly higher than the chemical element O II. The minimum temperature rise when the contact opening velocity 10 m/s, is the chemical element Cu II. This is not significantly different from the minimum temperature rise of the contact opening velocity 1 m/s, at the wavelength 776 nm of the same chemical element Cu II. The average temperature rise when the contact opening velocity 10 m/s, is 0.22×10^{-33} K higher than when the contact opening velocity 1 m/s.

6.8.3 Contact Material Ag/C step and Cu punch

There are 25 chemical elements when using the Ag/C step contact material, which has 6 chemical elements more than the Cu Punch contact material. At the wavelength 500 nm, the Ag/C step contact material shows the maximum electron energy which less than the Cu Punch contact material about 0.67×10^{-37} J. There is only one chemical element, O II appears in both Ag/C step and Cu Punch contact material. Both contact Material Ag/C step and Cu punch have the same lowest energy is at the wavelength 821 nm for chemical elements N I, O I and O II about 2.42×10^{-37} J. The average energy for the chemical elements of the Ag/C step contact material is less than when using the Cu Punch contact material about 0.89×10^{-37} J.

The Mass of the chemical elements Ag I and Cu II is higher than other chemical elements about 4 times in the Ag/C step contact material. The chemical element H has the lowest Mass when using Ag/C step contact material. The average Mass of Ag/C contact material is 0.134g. The maximum Mass when using the Cu Punch contact material is the chemical element Cu II at the wavelength 776 nm. The average Mass of the Cu punch contact material is 0.108 g. This confirms that the chemical element Ag I and Cu II come from the contact material. Both the Ag/C step contact material and Cu Punch contact material show the minimum Mass is the chemical element H.

The highest temperature rise of Ag/C step contact material is belong to chemical element N III at the wavelength 500 nm and the temperature rise of the chemical element O II and O IV are a little bit lower than the chemical element NIII. The maximum temperature rise for the Cu punch contact material is the chemical elements NII at the wavelength 428 nm, but the second highest is the chemical element O II, which it is nearly the same temperature rise as the chemical element N II. Also the wavelength at the maximum energy Cu punch contact material is shorter than the Ag/C step contact material.

Both Ag/C step contact material and Cu Punch contact material have the same lowest temperature rise and the same chemical element Cu II at the wavelength 776 nm. The average temperature rise for the Ag/C contact material is 2.49×10^{-33} K which is lower than Cu Punch contact material about 0.25×10^{-38} K. Therefore, temperature rise of Ag/C step contact material is less than the Cu punch contact material.

CHAPTER 7

DISCUSSION

7.1 Introduction

This chapter concentrates primarily on the effects of contact velocity and arc chamber venting on arc root mobility in the contact region. Experimental data from the optical system, pressure measurements and spectrometer measurements as well as modelling of the arc energetics and gas flow are used to analyse the effects of these parameters on the arc motion.

7.2 Review

A high speed Arc Imaging System (AIS) is used to record the arc root commutation from the contact region. A Flexible Test Apparatus is used to simulate the current limiting operation under controlled experimental conditions.

The optical data permits the identification of the cathode and anode arc roots on the fixed and moving contacts. Two pressure transducers are used to monitor the gas pressure in the arc chamber. A miniature optical spectrometer is used to measure the spectrum data.

The objective of the work here is to investigate the arc root contact time, pressure and element species in the arc chamber. The main investigations involve with the gap behind the moving contact, arc chamber venting, short circuit current level, contact opening velocity, arc chamber and contact material.

7.3 Influence of low contact opening velocity

7.3.1 Contact gap

The contact gap at the time that the arc root moves from the contact obtained from the experimental results in Section 4.3.4, Chapter 4, is shown in Figure 7.1. The contact gap increases as the contact opening velocity is increased. The contact gap for the anode root is higher than the cathode root.

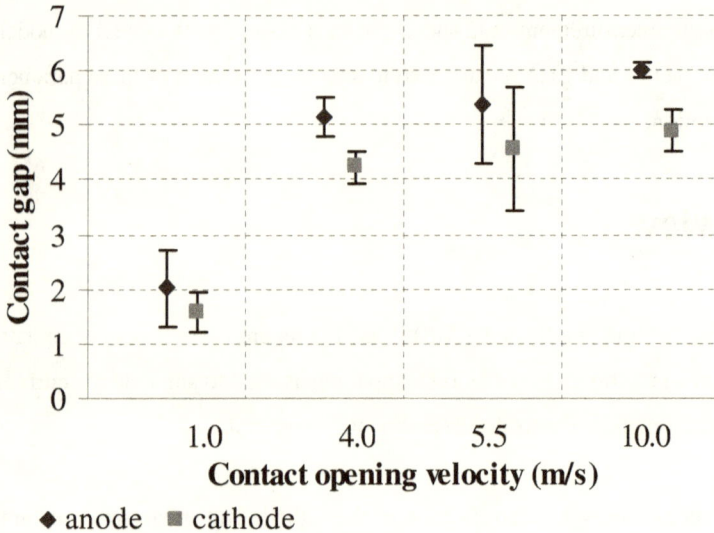

Figure 7.1: Contact opening velocity and contact gap

The results here shows that at low contact opening velocity, the critical contact gap that the arc root starts to move from the contact region is less than that reported by Rieder [35], when observing the minimum gap that the arc starts to move from the contact region.

Rieder's experimental conditions were an arc current of 2 kAdc and a magnetic blast field 25mT/kA. The arc always commutated, independent of the material, for a contact gap of approximately 2 mm. This is called minimum contact gap by Rieder .

The new results here disagree with [35] showing that at a low contact opening velocity of 1 m/s, the arc root moves from the contact region before the contact gap reaches 2 mm.

The reason for this difference lies in the test system. In this work, the short circuit arc current was set at 2 kAac, with the arc being created by the contact opening system. The arc chamber geometry used a straight electrode for the fixed contact and a 45 degree electrode for the moving arc runner.

In the AC test system, the arc current was a function of time as was the contact opening velocity. Hence, the arc current at the point that the arc occurred was lower than at the point that the arc root moved from the contact region.

Rieder [35] used parallel electrodes in fixed positions in the DC test system. The arc was ignited by a blast field circuit. The arc current was then constant for the whole operation from the period that arc occurred until the arc was extinguished. At the point that the arc root moved from the contact region the experimental results here obtain both the arc current and the contact gap at various contact opening velocities.

The experimental results here agree with [35] in that the arc root moves from the contact region when the contact gap is more than 2 mm when the contact opening velocity is significant more than 1 m/s (eg. 4 m/s). Hirose [49] also confirmed that when the contact velocity was greater than 0.3 m/s the critical gap was less than 5 mm.

The critical contact gap was found to be related to the arc current, the magnetic field and the contact opening velocity. An optical system was used to record the arc position on the rotating drum. The actual tests were carried out with arc current of 8.6 kA and 1.5 kVdc.

The new results show that although the contact gap (1.57 mm) is less than the minimum stated in [35], the arc can still moves off from the contact region. This confirms that the arc root contact time is effected by the contact gap and the contact opening velocity. This analysis suggests that the contact gap is not the only factor in determining whether the arc root will move from the contact at low contact opening velocity.

This work provides a contribution of knowledge by demonstrating that at low contact opening velocity the arc starts to move off from the contact region with a contact gap less than previously investigated [35].

This work aims were achieved to the extent that the AIS and FTA were used to investigate the influence of low contact opening velocity on the anode and cathode root motion from the contact region.

7.3.2 Arc root contact time

The new results in Section 4.3.4, chapter 4, show that the arc root contact time decreases as the contact opening velocity is increased. This does not completely agree with the observations of Belbel [15] who considered the arc motion with contact opening velocity.

The arc root contact time that Belbel observed was dependent on the contact opening velocity when the contact opening velocity was lower than 6 m/s. For contact opening velocities higher than 6 m/s, the arc root contact time was independent on the contact velocity and the contact time is minimized.

The results here are in agreement with Belbel [15] in that the arc root contact decreases as the contact opening velocity is increased for contact opening velocity from 2 m/s to 6 m/s. This is also supported by the results of Behrens [74] who used a single optical sensor to measure the arc contact time. Increasing the contact velocity and magnetic flux density resulted in lower arc contact time.

The results here disagree with [15] in that the arc root contact time is dependent on the contact opening velocity for contact opening velocities lower than 6 m/s. The results of this work show that the arc root contact time decreases as the contact opening velocity is increased for contact opening velocities from 1 m/s up to 10 m/s.

The reason for the difference is that Belbel [11] observed the influence on the arc motion of contact opening velocities from 2 m/s to 13 m/s by using an optical sensor in a parallel conductor arc chamber, with a short circuit arc current of approximately 10 kA.

Meanwhile, this work used the AIS to record the optical data, with the contact opening velocity fixed in the range from 1 m/s to 10 m/s. The cathode was on the fixed contact and the short circuit arc current was set at 2000 A.

With contact opening velocities over 6 m/s, the results from [15] reported that the arc root contact time hardly changed, however, the new results here show clearly that the arc root contact time decreases as the contact opening velocity is increased. It could be that when the contact opening velocity was higher than 6 m/s, the arc moved very fast.

Thus, it was very difficult to measure the difference of the arc root contact time with Belbel's [15] limited equipment. The arc root contact time that Belbel [15] used was in the scale of milliseconds but the new AIS equipment used here captures the optical data every 1 µs.

This provides benefits to observing the arc motion at a smaller scale. From the experimental results in Figure 4.10, section 4.3.4, the arc root contact time for contact opening velocities between 5.5 m/s and 10 m/s was different by about 300 µs. The limitation of the study in [15] on the effect of the contact opening velocity on the arc motion was that the equipment was not fast enough to capture the event.

Belbel [15] also reported that when the contact opening velocity higher than 6 m/s, the arc root contact time is minimized. The results in Section 4.2 to section 4.4, show that at reduced contact opening velocity down to 1 m/s, the arc root contact time can be the same value as high contact velocity of 10 m/s. This can be obtained by limiting the vent between the moving contact and the arc runner, the effective contact material and arc chamber material as shown in Figure 7.2.

The condition of the high contact opening velocity of 10 m/s is Ag/C contact material, ceramic arc chamber and opened the gap behind the moving contact. Meanwhile, the condition for low contact opening velocity of 1 m/s is Cu punch, Polycarbonate and closed the gap. Figure 7.2 shows the relationship between the arc root contact time and variable parameter.

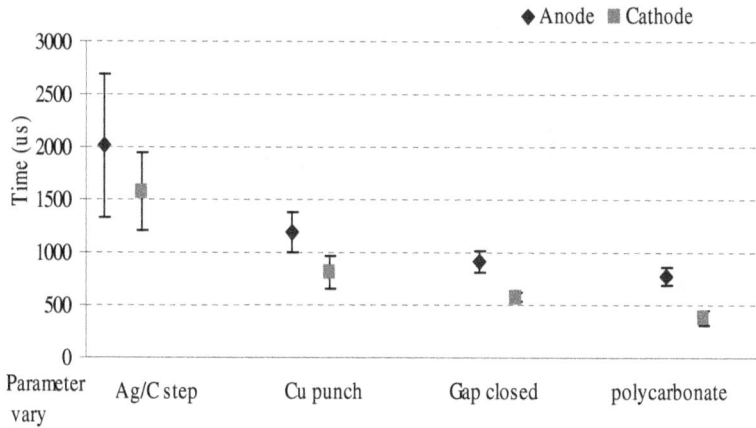

Figure 7.2: Arc control for low contact opening velocity

This work provides a contribution of knowledge in that the high speed AIS, which was able to record optical data of arc motion at sample rates of 1 MHz, showed that the arc root delay time was clearly dependent on the contact opening velocity. These new results provide advantage more than the previous observation [15].

This work aims were achieved to the extent that the high speed AIS provided a benefit to closer investigation of the influence of the contact opening velocity on the arc root motion from contact region.

7.3.3 Arc current

The results here are demonstrated that at the point that the arc root moves from the contact region, the arc current is dependant on the contact opening velocity from 1 m/s to 10 m/s as shown in Figure 7.3.

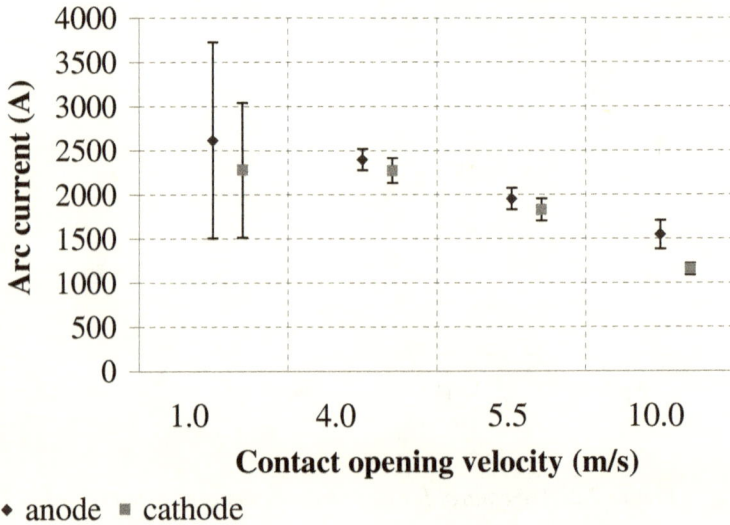

Figure 7.3: Contact opening velocity and arc current at the time that arc root moves off from the contact regions, Ag/C step, ceramic arc chamber, choked arc chamber venting

At the point at which the arc root moves from contact region, the short circuit arc current increases as the contact opening velocity is decreased. The results here are supported by the investigations of Widmann [39]. The arc commutation delay decreases with increasing arc current. In his experiments, the values of arc current and the commutation delays on the anode and cathode root were measured.

The results here also confirm the experimental results of Guile [75] in that the arc current affects the arc motion by changing the conditions in the cathode spot. A contact gap of 3.2 mm, with a magnetic flux density 0.027 Wb/m^2, was used in his experiments.

This suggests that at the point at which the arc moves from the contact region when the contact opening velocity is increased, the arc current and the arc root contact time decreases. When the short circuit current increases, this induces an increase in the electromagnetic field.

7.3.4 Magnetic forces

From the magnetic force modeling in Section 5.3, Chapter 5, the results of the modelling with a variety of contact opening velocities are shown in Figure 7.4. The results show that the magnetic forces per arc length decrease as the contact opening velocity is increased.

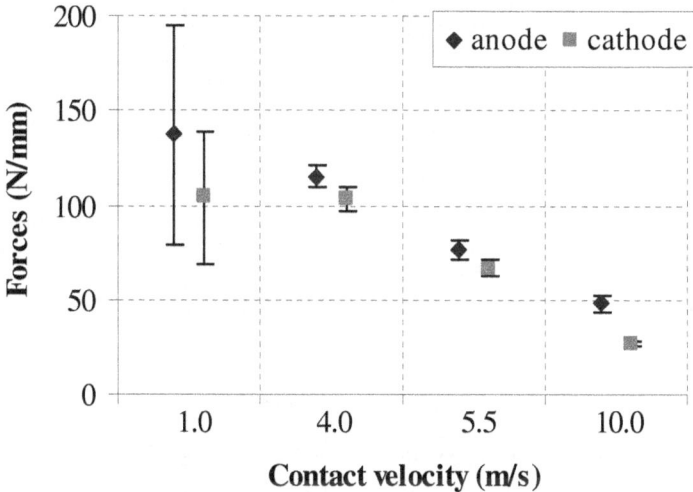

Figure 7.4: The force per arc length at the time that the arc root moves off from the contact regions

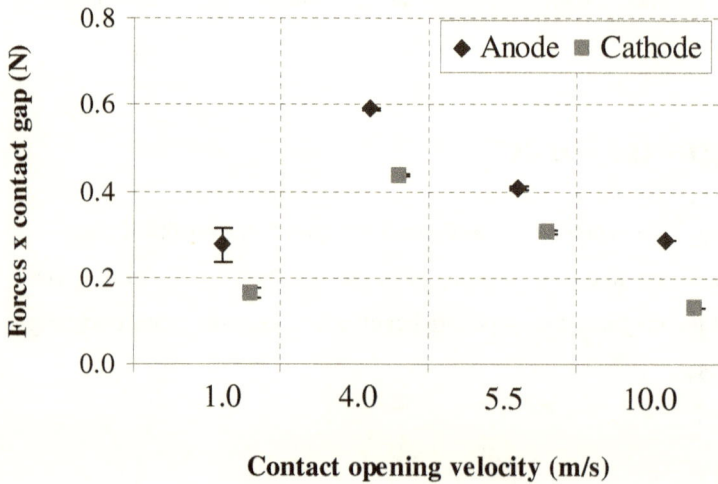

Figure 7.5: The total force at the time that the arc root moves off from the contact regions

The total forces (approximately) integrated along the arc length and contact opening velocities are shown in Figure 7.4. The results show no simple trend with contact opening velocity. However, the data are compatible with an approximately constant total force.

The results of the magnetic forces modeling here are lower than the modeling results of Paul [11] who considered the magnetic driving force in the contact region from the magnetic field from the conductors and steel plates in the side walls. The magnetic flux density (B) was 31 mT/kA for the conductors in the air with a gap of 10 mm. Thus, at an arc current 2000 A, the magnetic forces was about 1.24 N.

The arc chamber geometry is considered. The magnetic flux density calculation here is considered at the point that the arc root moves from the contact region with contact opening velocities.

The arc current and the contact opening velocity are a function of time and obtained from experimental results. The arc current is determined at the time that the arc root moves from the contact region. The width of contact gap depends on the contact opening velocity.

The experimental results here show the maximum contact gap is less than 10 mm. The magnetic forces at the point that the arc moves from the contact region are approximate 0.2 N for contact opening velocities from 1 m/s to 10 m/s.

The results of the magnetic forces modeling here agree with the value of the magnetic blast-field (B) that Rieder [35] used in the experiments to investigate the arc motion across the gap and step. The experimental conditions that Rider [37] used were an arc current of 2 kAdc. The magnetic blast field 25 mT/kA was applied to force the arc to move from the contact region across the contact gap of 2 mm.

The results here also agree with Guile [76] in that the magnetic field decreases with increased distance from the electrodes. Guile studied the movement of an AC arc with a wind opposing the electromagnetic forces by using parallel mild steel electrodes. The arc current was up to 3000 A.

This finding showed that the critical magnetic forces for the arc to move from the contact region with contact opening velocities between 1 m/s and 10 m/s was about 0.2 N for a short circuit arc current of 2000 A and 340 V. This suggests that the results here show that both the contact gap and arc current have an influence on the magnetic forces.

This works aims were achieved to the extent that magnetic force modeling was used to contribute to a solution of the behaviour of the arc root commutation from the contact region at reduced contact opening velocities. The contribution to knowledge is the critical magnetic forces that the arc root starts to move from the contact region is independent on the contact opening velocity.

7.3.5 Arc root velocity

The results from Section 5.3, Chapter 5 show that the velocity of the gas flow is higher than the arc root velocity. At the point that the cathode root moves from the fixed contact, the arc velocity is about 1.6 m/s and gas flow velocity is about 4.4 m/s. For the anode to move from the moving contact, the arc velocity is about 25.4 m/s and gas flow velocity is about 31.78 m/s.

This is supported by the experimental and modeling results of Paul [11] who reported that the arc velocity was about 20 m/s and the flow velocity was 32 m/s when the arc moved off from contact region. Only the velocity of the anode root in this work is compatible with the experimental results in [11].

The results here implicitly prove that the arc root velocity can be observed for the anode and cathode root individually, with the velocity of the anode root being higher than the cathode root. The velocity of the cathode root is a much lower value than specify in [11].

Paul used a miniature circuit breaker of conventional current limiting design. The flow velocity was calculated from the one-dimensional shock wave by using the pressure ratio. The arc root velocity was obtained from optical measurement.

The reason for this difference is thought to lie in the experiment methodology. The flow velocity in this work was approximated from the gas dynamic model by using the ratio of the pressure across the arc. The arc root velocity was estimated from the optical data as described in section 5.3, chapter 5.

These relations describe the gas flow in front of the arc as the arc moves away from the contacts towards the arc chamber. The arc is driven by magnetic forces as described in Chapter 1, but also by pressure build up behind the arc caused by heat and gas flow from the arc into the restricted space behind the contacts.

The arc trajectories in [11] were obtained by calculating the centre of the light intensity distribution. The shape of the arc in this work used five dynamic threshold levels to compute contour levels of the light intensity distribution. This provides the advantage that the arc contour images can be generated at low light levels during the early stages of the arc event.

This is useful to gain knowledge of the arc root motion from the contact region. In addition, the arc root velocity here is observed for both the anode and cathode root individually.

The flow velocity in [11] was calculated from the balance between magnetic and gas dynamic forces. The modeling of the magnetic flux density (B) was used to predict the pressure across the arc. The flow velocity was calculated from the gas behind the shock wave. The flow velocity estimation here, the pressure ratio, is obtained from experimental results.

This gives the advantage that the pressure data can be obtained from the individual arc root. The flow velocity at the point that the anode and cathode root move from the contact region can be approximated. In this work the arc root velocity and flow velocity investigated here can be estimated on both the anode and cathode root move from the contact region.

The findings here show clearly that at the point at which the cathode and anode move from the contact region, the velocity of the gas flow is higher than the arc root velocity. This can be explained by the arc heating and melting the contact as it stays in the contact region, leading to a high temperature and pressure in the arc chamber. The high temperature from heating the contact in that area also has an effect on the gas flow in the arc chamber.

From the relationship between arc velocity and the magnetic forces, the work of the arc root displacement can be approximated. The magnetic forces in section 7.3.5 show about 0.2 N for contact opening velocities between 1 m/s and 10 m/s. Therefore, at the point that the arc root moves from the contact region, the work of the arc root displacement is approximately 0.32-0.88 W for the cathode root and 5.08-6.36 W for the anode root.

This work aims were achieved to the extent that the combination of the magnetic and gas dynamic forces in the contact area was used to calculate the work of the arc root commutation from the contact region at reduced contact opening velocity. The contribution to knowledge is the anode root velocity on the moving contact is more than the cathode root on the fixed contact.

7.3.6 Arc power

At the point that the arc root moves from the contact region, the average arc power for contact opening velocities from 1 m/s to 10 m/s is considered constant as shown in Figure 7.6. There is little variation in arc power with contact opening velocity.

The arc power is a function of arc current and arc voltage as well as arc root contact time. At low contact velocity the arc voltage at the point at which the arc root moves from the contact region is lower than at high contact opening velocity but the arc current is higher. Therefore, the arc power at the point at which the arc root moves from the contact region on the different contact opening velocity are in the same range.

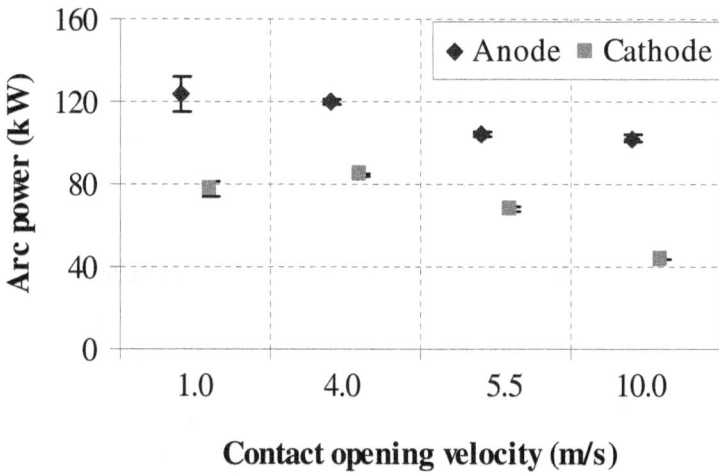

Figure 7.6: The arc power at the time that the arc root moves off from the contact regions and contact opening velocity

Previous work has indicated a minimum contact gap required for arc motion, although the physical mechanism for this is not clear [11,37]. This finding suggests that a minimum arc power is required for arc motion. The discussions above show how the arc power is related to the thermal and fluid dynamic processes in the arc chamber. The minimum arc power requirement can therefore be used to obtain a more direct understanding of the underlying physics controlling arc immobility.

From section 7.3.5, the work for the arc root movement was very little. It was assumed that the arc power transforms to the work of the arc root displacement and the heat that occurs in the contacts and the hot gases in the contact area. Therefore, most of the arc power is converted to heat.

The findings show that the arc power at the point at which the arc root moves from the contact region appears to have no significant linkage to the contact opening velocity between 1 m/s to 10 m/s.

7.3.7 Mass flow

The mass flow calculation provides a very top level estimation of the volume of gas being produced in the arc chamber. This gas must flow through the arc chamber vent, this is possible to estimate the rate of flow and the flow velocity.

This could be a starting point to model the gas composition in the arc chamber. Initially the flow will be air, but this would be used up and be replaced by silver from the contact material. As the arc starts to run, this would start to produce element from the steel runners and splitter plates.

The flow velocity can be related to the measured values of pressure drop in the arc chamber. The thermodynamic calculation provides the flow rate produced by the evolution and expansion of the gas. There is also a flow produced by the movement of the arc due to magnetic forces. By comparing the thermodynamic model with the measurements, this could estimate the contribution from the electrodynamic forces.

The thermodynamic calculation presents in this book is an early stage to look at the overall energetic of the circuit breaker operation. This provides a powerful tool for building in with electrodynamic and gas dynamic models to approach the arc modeling. This is for a future project

The mass flow rate and the total mass flow at the point that the arc root moves from the contact region are shown for a variety of contact opening velocities in Figures 7.7 and Figure 7.8.

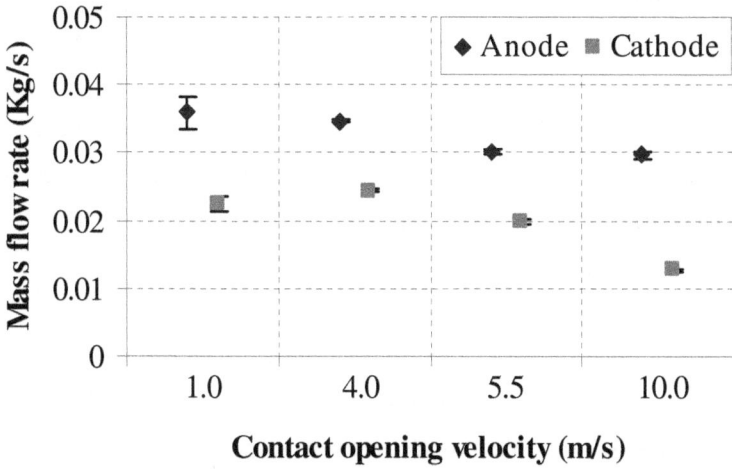

Figure 7.7: *The mass flow rate at the time that the arc root moves off from the contact regions and contact opening velocity*

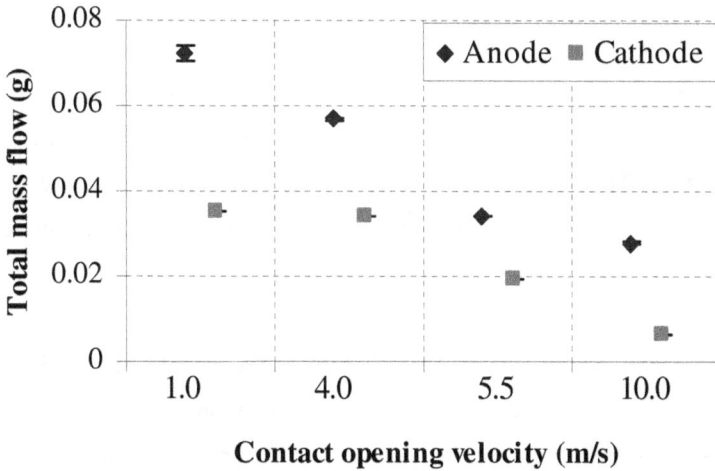

Figure 7.8: *The mass flow at the time that the arc root moves off from the contact regions and contact opening velocity*

The total mass flow at the point that the arc root moves from the contact region shows that the mass flow decreases as the contact opening velocity is increased.

The results here confirm McBride [77] in that increasing the contact opening velocity reduces the material transfer. The contact opening velocity has a significant effect on the erosion of the contacts. The material transfer will be from cathode to anode (cathodic erosion) at low arc current.

At high arc current, the material transfer will be from anode to cathode (anodic erosion). McBride [77] investigated the AC erosion with contact opening velocities from 0.1 m/s to 0.8 m/s. The maximum contact gap was 10 mm. The supply voltage was set at 240 V, with currents of 5.6 A, 14.7 A and 26.3 A. When the cathode was on the moving contact, the cathode mass loss (negative erosion) and the anode mass gain (positive erosion) was always less.

The work here provides more information about material transfer with varying contact opening velocity than the previous investigation in [77]. This work presents the mass flow for both the anode and cathode root with contact opening velocities from 1 m/s to 10 m/s. The limitations of the study in [77] were that the equipment had limited control of the velocity of the contact up to 0.8 m/s and that the arc current was less than 100 A.

The results of this work are supported by the experimental results of Chen and Sawa [78] who investigated the effect of arc behaviour on material transfer in a breaking arc. They presented that when the arc current increased, the anode gain and the cathode loss increased.

The new results here suggest that the mass flow and the arc current decreases as the contact opening velocity is increased. The material transfer will be from cathode to anode. In low contact opening velocity, the mass flow and the arc current are increased. The material transfer will be from anode to cathode [77].

The mass flow from the cathode to anode decreases at high contact opening velocity. The mass flow from the anode to cathode increases at low contact opening velocity. The contribution to knowledge is that the total mass flow decreases as the contact opening velocity is increased.

7.3.8 Spectrum

From the results in Section 4.5, Chapter 5, a higher quantity of electronegative species is observed at higher contact velocities. The element species of oxygen and nitrogen appear only at the high contact opening velocity.

The emission of the arc chamber walls is in agreement with that observed by Sato and others [47,48,52,53]. The arc spectral is emitted from the contact and arc chamber material. A large amount of the arc chamber walls molecules, such as Hydrogen and a small fraction of carbon, are vaporized into the plasma.

The results here show clearly the spectral of Ag when an Ag/C step was used as a contact material. This is in agreement with Kiyoshi and others [42,57,54]. The spectrum of Ag is at the wavelength 521 nm and 546 nm and is visible in both ceramic and polymer arc chambers with Ag/C contact material.

This is also supported by the investigation of the mass flow in section 7.3.7. The amount of the metallic vapour and gases flow in the arc chamber with low contact opening is higher than that of high contact velocity.

The findings here show that the composition of the arc gases is shown to depend on the contact opening velocity. The spectral emission from the arc gases confirms the presence of the electronegative species from the polymer arc chamber walls.

This suggested that at reduced contact opening velocity, the mass flow in the arc chamber increases. Hence, the metallic vapor flows out from the arc chamber via the vent between the moving contact and the arc runner. Therefore, there is less element species left in the arc chamber.

In addition, for low contact opening velocity the arc roots stay on the contact region longer, see section 7.3.2. This melts the contact and heats up the polycarbonate arc chamber. This chemical process uses more oxygen and nitrogen. This work aims were achieved to the extent that the arc gas composition in the arc chamber was observed using a spectrometer.

This work provides a contribution of knowledge by demonstrating that the elements of nitrogen and oxygen appear more at high contact opening velocities than at lower velocities.

7.4 Influence of the venting

7.4.1 Area venting

The larger area venting gives a lower arc root contact time. The influence of the arc chamber venting has an effect on the arc root of the moving contact as shown in the experimental results in Figure 4.5, section 4.3.2, Chapter 4.

The relationship between the vent area and arc root contact time is shown in Figure 7.9. It shows that the vent area of the arc chamber is critical to the arc root's motion away from the contact region.

The area venting has a strong effect on the arc root contact time of the moving contact. The venting arrangement here has a stronger influence on the cathode root motion than the anode root.

The stronger effect of the area venting is on the cathode root motion. The lower cathode root contact time on the moving contact is contrary to the observation of Widmann [39]. The cathode root delay time for crossing a gap has been observed to be longer than for the anode root.

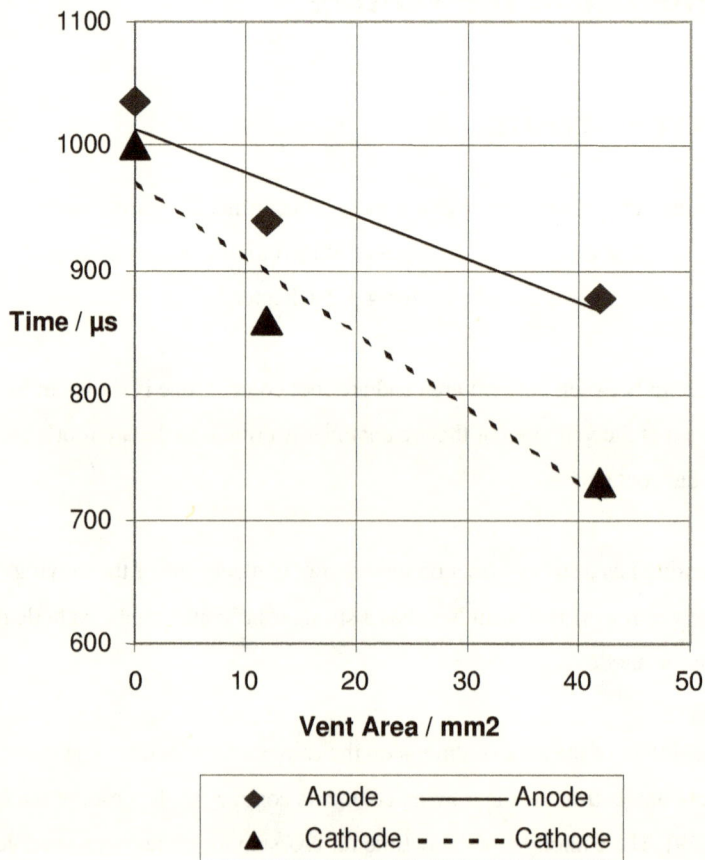

Figure 7.9: Effect of vent area on moving contact and arc root contact times

Widmann observed the delay of cathode and anode roots over various size steps and gaps in conductors. It was shown that the delay with steps or gaps geometry was greater for the cathode root. For the gap widths 1 mm to 5 mm, the cathode delay was approximately 10 times longer than the anode.

The new finding here show that as the vent area is reduced towards zero the anode and cathode root contact times become a similar value. As the vent area is increased, the cathode root contact time is lower than the anode root.

The reason for this difference is thought to lie on the arc chamber geometry with the open gap between the moving contact and the arc runner. The experimental conditions in this work have the arc drawn between opening contacts. The arc runner in the fixed contact is 90 degrees and 45 degrees for the moving contact arc runner.

The gap between the moving contact and the arc runner is 2 mm wide and opens directly to air outside the arc chamber. It is considered that the open gap in the arc chamber caused this difference. Widmann's observations [39] were with an arc ignited between parallel runners with a gap depth of up to 7 mm.

This is in agreement with Shea and others [17,38]. The arc root motion is dependant on the quantity of the flow and area of the arc chamber venting. They investigated the influence of the area venting with a contact opening velocity of 4.1 m/s and contact gap of 1.6 cm. Increasing the venting area of the arc chamber causes an increase in wall vaporization, arc current and flow.

As the vent area was increased, the arc root contact times decreased. The vent area determined the pressure developed during an interruption and the flow of hot gas and metal vapor from the arc chamber.

The findings show that the velocities of the opening contact have an effect on the arc root motion across the gap. In addition, with an opened gap the thermally driven flows, from the emission of metallic vapor from the contact region and the hot gases, flow away from the arc and out of this gap.

7.4.2 The gap behind the moving contact

The gap behind the moving contact creates a reduced pressure in this region. The resulting adverse gas flow impedes the arc root commutation. This results in an observed increase in arc root contact times compared to a closed gap. The gap behind the moving contact has a strong effect on the arc root on the moving contact. The delay time of the arc root is 17-24 % less than with the gap opened as shown in Section 4.3.5, Chapter 4.

The new results show that when the arc crosses the opened gap, the anode root delay is longer than the cathode. While the arc is crossing the depth gap, the cathode root delay is longer than the anode [39]. The reason for this difference is though to lie in the gas flow conditions within the arc chamber, particularly around the moving contact.

This flow pattern is impedes transfer of the anode from the moving contact to a greater extent than the cathode. The anodic gases and deionises are swept away in the opened gap between the moving contact and the arc runner. This reduces energy transfer to the moving contact arc runner. It prevents the extension of the conductive arc column and the establishment of a new electron receptor site. As discussed in [39], the anode delay may be more impeded than the cathode.

The cathode is associated with a large volume of ionized material electrostatic drawn to the cathode region. The lower cathode mobility and higher cathode power dissipation cause the generation of the relatively high quantities of the vapourised material.

This will be entrained into the gas flow behind the moving contact to establish a conductive area behind the moving contact. Material will also be directed towards the arc runner by plasma jets, which are prevalent on the cathode. Both these effects combine to promote conditions for establishing a new arc root on the moving contact.

This shows an agreement with Widmann and Takenaka [39,45]; the gas flow in the region of the moving contact is proposed as the mechanism resulting in the delay of the arc root. This is also supported by the modeling of the gas expansion in the simple three layers and optical measurements [79]. It was reported that the arc starts to move when enough highly ionised gas was in the arc chamber.

This suggests that the geometry of the arc chamber of the commercial MCBs with an open gap between the moving contact and the arc runner can be improved. The movement of the arc root from the contact region can be enhanced by limiting this gap.

This work provides a contribution of knowledge by demonstrating that area venting has a strong effect on the cathode root movement across the open gap on the moving contact. This observation contrasts the previous investigation [39].

When the arc root arrives at the edge of the moving contact. The flow in front of the arc would be channelled down the contact edge and flow into the gap between the moving contact and the arc runner.

This results in a low mass of anodic gases obstructing the transfer from the moving contact to the arc runner. The amount of the mass flow when the arc has not left the contact region is shown in section 7.3.8. Thus, the motion of the arc root on the moving contact would be slowed down.

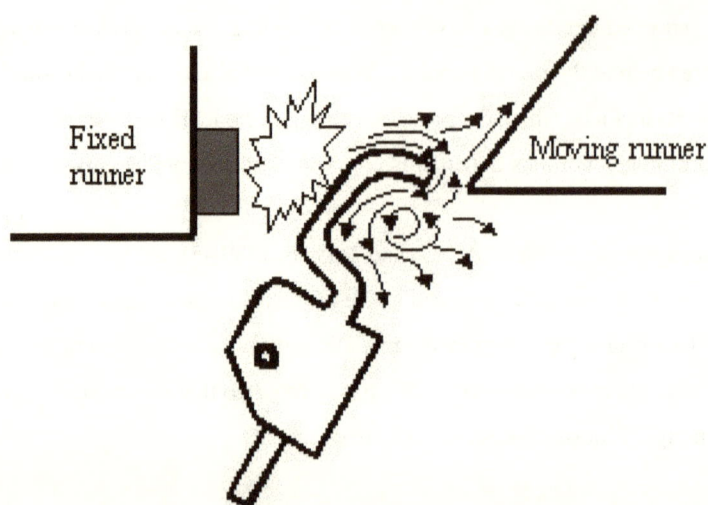

(a)

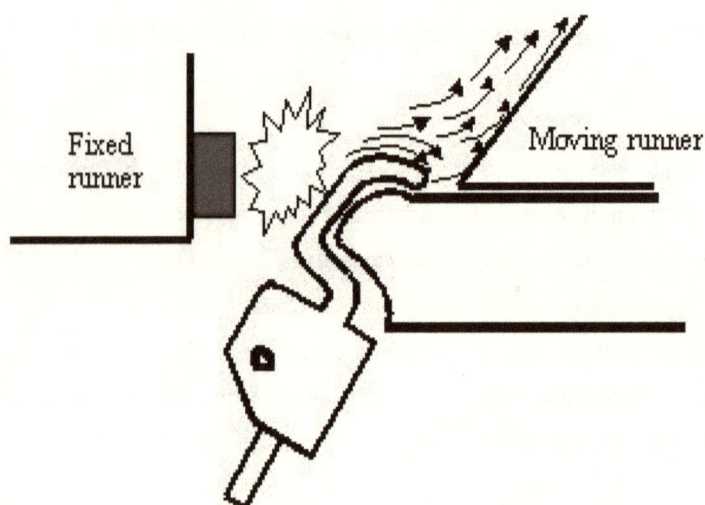

(b)

Figure 7.10: Gas flow around the moving contact as the gap opened (a) and closed (b), ceramic arc chamber material, Cu punch contact material, contact opening velocity 10 m/s, peak short circuit current 2000 A, choked arc chamber vent and supply current polarity as anode root on the moving contact and cathode root on the fixed contact

7.5 Energy and Temperature rise

7.5.1 Polycarbonate and Ceramic Arc chamber

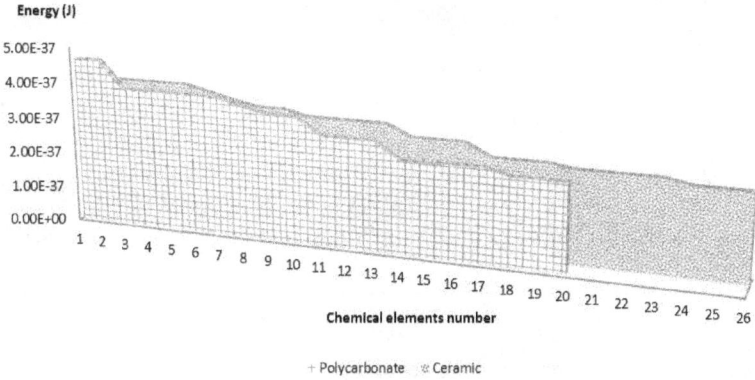

Energy (J) vs Chemical elements number
+ Polycarbonate ☆ Ceramic

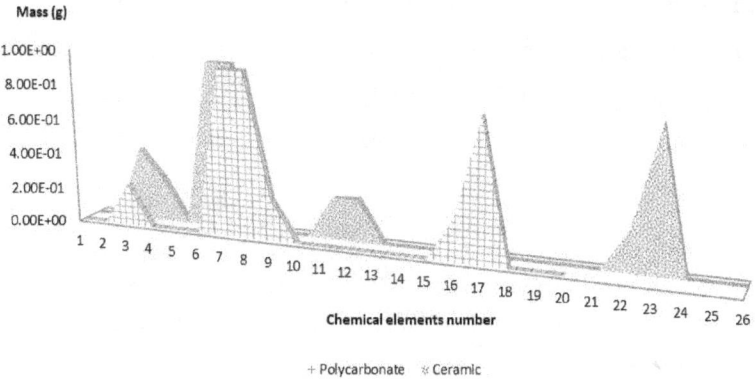

Mass (g) vs Chemical elements number
+ Polycarbonate ☆ Ceramic

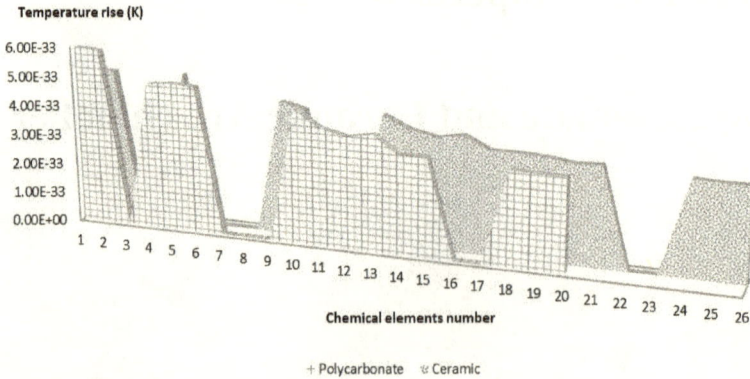

Figure 7.11:Energy, Mass and Temperature rise in Polycarbonate and Ceramic arc chamber

Both the polycarbonate arc chamber and the ceramic arc chamber show that the electron energy is decreased as the wavelength increases. The maximum energy polycarbonate arc chamber is higher than the ceramic arc chamber, but overall energy in the ceramic is more than the polycarbonate arc chamber. Both the polycarbonate arc chamber and ceramic arc chamber have the same minimum energy and have the same chemical elements N I, O I and O II. The polycarbonate arc chamber and ceramic arc chamber have the same chemical elements at the maximum energy which it is O II.

The maximum Mass in the Polycarbonate arc chamber is as same as the ceramic arc chamber. This Mass comes from the chemical element Ag I which it is the main ingredient of the contact material. The chemical element H is the minimum Mass in both the Polycarbonate and the ceramic arc chamber.

The maximum temperature rise in the polycarbonate arc chamber is higher than the ceramic arc chamber. From the same chemical element Cu II, the minimum temperature rise of the polycarbonate arc chamber is more than the ceramic arc chamber. The average temperature rise in the polycarbonate arc chamber is higher than the ceramic arc chamber.

7.5.2 Contact Opening Velocity 10 and 1 m/s

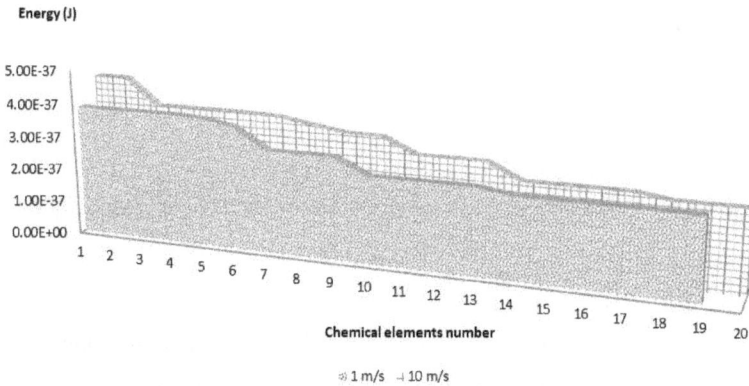

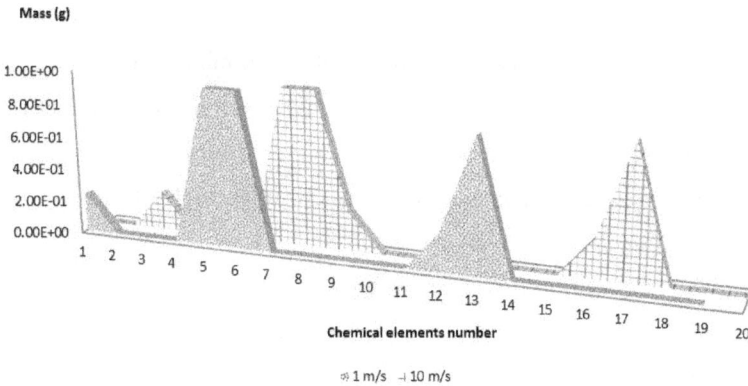

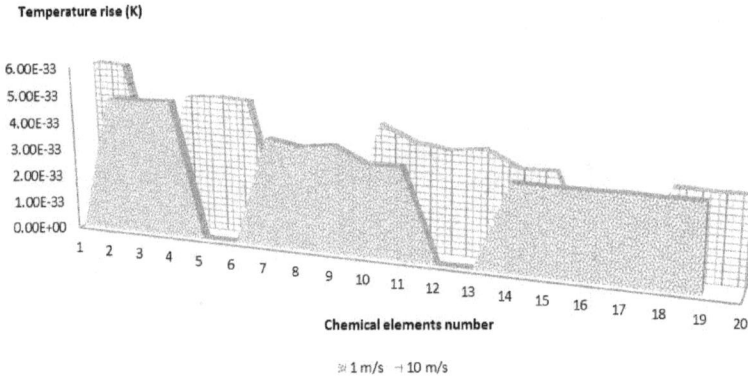

Temperature rise (K)

6.00E-33
5.00E-33
4.00E-33
3.00E-33
2.00E-33
1.00E-33
0.00E+00

1 2 3 4 5 6 7 8 9 10 11 12 13 14 15 16 17 18 19 20

Chemical elements number

≥ 1 m/s   ┤ 10 m/s

Figure 7.12: Energy, Mass and Temperature rise when contact opening velocity 1 m/s and 10 m/s

The maximum energy when contact opening velocity at 10 m/s is higher than the contact opening velocity 1 m/s. However, they have the same minimum energy and the same chemical element N I, O I and O II. The average energy when the contact opening velocity 10 m/s is slightly higher than the contact opening velocity 1m/s.

Both contact opening velocity of 1 m/s and 10 m/s has the same minimum and maximum Mass. The maximum Mass, when the contact opening velocity 10 and 1 m/s is the chemical element Ag I. The minimum Mass is the chemical element H. The Mass average when the contact opening velocity 10 m/s is 0.003g higher than the contact opening velocity 1 m/s.

The maximum temperature rise when the contact opening velocity 10 m/s, the chemical element N II, is higher than the contact opening velocity 1m/s the chemical element N V. The minimum temperature rise of the contact opening velocity 10 m/s, is the chemical element Cu II, is not significantly different from the minimum temperature rise of the contact opening velocity 1 m/s. The average temperature rise when the contact opening velocity 10 m/s is 0.22×10^{-33} K higher than when the contact opening velocity 1 m/s.

7.5.3 Ag/C step and Cu punch contact material

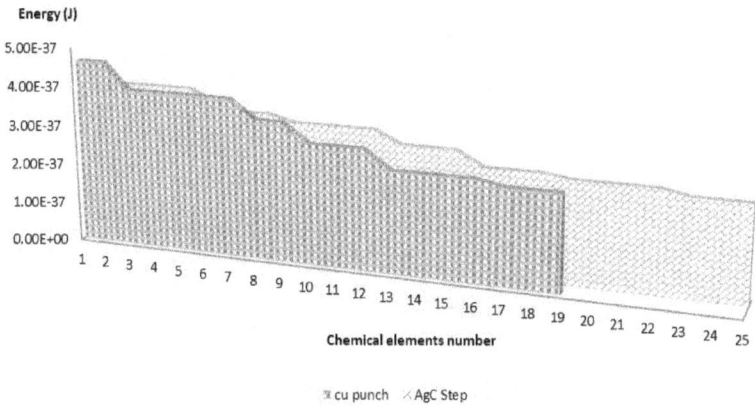

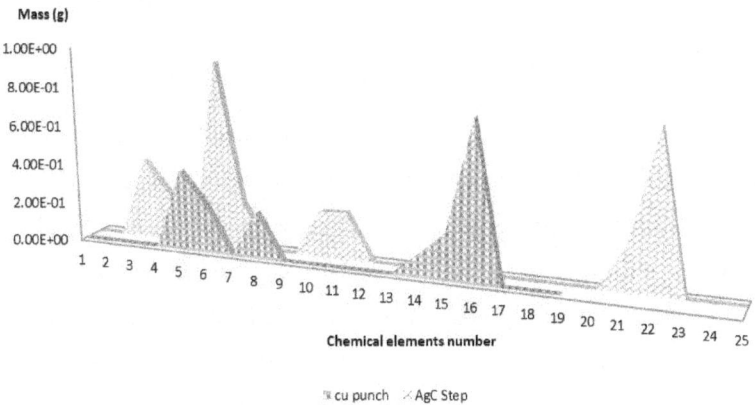

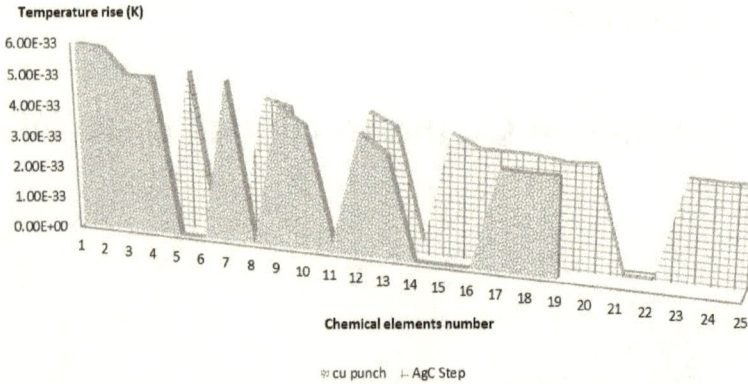

Figure 7.13:Energy, Mass and Temperature rise for Ag/C step and Cu punch contact material

The Ag/C step contact material shows the maximum electron energy less than the Cu Punch contact material. However, there is only one chemical element, O II appears in both Ag/C step and Cu Punch contact material. Both contact Material Ag/C step and Cu punch have the same lowest energy at the wavelength 821 nm for chemical elements N I, O I and O II.

The Mass of the chemical elements Ag I and Cu II is higher than other chemical elements about 4 times in the Ag/C step contact material. The maximum Mass when using the Cu Punch contact material is the chemical element Cu II at the wavelength 776 nm. This confirms that the chemical element Ag I and Cu II come from the contact material. Both the Ag/C step contact material and Cu Punch contact material show the minimum Mass is the chemical element H. The average Mass of Ag/C contact material is 0.026g higher than Cu Punch contact material.

The highest temperature rise of Ag/C step contact material belongs to chemical element N III at the wavelength 500 nm. The maximum temperature rise for the Cu punch contact material is the chemical elements NII at the wavelength 428 nm. The maximum temperature rise for the Cu punch contact material is higher than the Ag/C step contact material. Both Ag/C step contact material and Cu Punch contact material have the same minimum temperature rise and the same chemical element Cu II, at the wavelength 776 nm. The average temperature rise for the Ag/C contact material is lower than the Cu Punch contact material about 0.25×10^{-33} K.

7.6 Summary

The AIS and FTA provide a benefit to closer investigation of the influence of the contact opening velocity on the arc root motion from the contact region. These new results show more implicitly than the previous observation [15].

The arc starts to move off from the contact region when the contact gap is less than previously thought [35]. The arc current is dependant on the contact opening velocity. The anode root velocity on the moving contact is more than the cathode root on the fixed contact. The total mass flow decreases as the contact opening velocity is increased. There are more electronegative elements more at high contact opening velocities than at lower velocities.

The magnetic forces are independent of the contact opening velocity. The arc power has no significant reducing the contact opening velocity. The area venting has a strong effect on the cathode root movement across the open gap on the moving contact. This observation contrasts with a previous investigation [39].

The combination of the magnetic and gas dynamic forces in the contact area was used to explain the arc root commutation from the contact region at reduced contact opening velocity.

The major mass in the contact area comes from the chemical elements from the contact material. The high opening velocity has a high energy, high mass and high temperature rise.

This work suggests that low contact velocity MCBs can gain the same performance of the arc root motion from the contact region as fast contact opening. It also makes the use of smart material actuation possible in these devices.

CHAPTER 8

CONCLUSIONS

8.1 Review

The arc root motion from the contact region in Miniature Circuit Breakers (MCBs) has been investigated in this book with the Flexible Test Apparatus (FTA) being used to simulate the operation of a miniature circuit breaker when a short circuit current occurs.

The Arc imaging System (AIS) was employed to record the optical data from the arc chamber in the FTA with two piezo resistive pressure transducers monitoring the gas pressure in the arc chamber and an optical fibre spectrometer installed to measure the spectral data during arcing.

This was the first time that this combination of the optical data, the pressure data and the arc spectrum was used to monitor and analyse the movement of the arc root from the contact region, allowing simultaneous observation of the gas flow and element species in the arc chamber.

The arc root contact time, pressure and arc spectrum in the arc chamber were investigated with contact opening velocities from 1 m/s to 10 m/s. The main investigations looked at the influence of:

- arc chamber venting;

- short circuit current level;

- contact opening velocity;

- arc chamber material;

- contact material;

- the gap behind the moving contact

8.2 Conclusion from the research

It is well known that the arc chamber vent has an effect on the arc root contact time. New results shown here demonstrate that the gap behind the moving contact also has strong effect on the arc root motion; this result being quantified at low contact velocity. Moreover, it is the first time that estimates of mass flow and volume flow of the arc gas that escape the gap behind the moving contact have been calculated.

New results from this work were used to estimate the minimum forces that moved the arc root from the contact region. The minimum force for the arc to move from the contact region, with contact opening velocities between 1 m/s and 10 m/s, was approximately 0.2 N. This value is however, only an estimate and further study is requirement to improve the accuracy.

It is known that the magnetic forces dominate the arc motion in the arc chamber, however, new results show that at the point at which the arc root moves from the contact region, the driven flow could dominate the arc root commutation from the contact region.

There have been investigations [15,35] that suggest that the arc root contact time is decreased as the contact gap increases for contact velocities below 2.2 m/s. When the contact velocity was more than 6 m/s, the arc contact time was a minimum. It has also been observed that the minimum contact gap for the arc root to move from the contact region was approximately 2 mm.

New results presented here show that:

- The contact gap increases as the arc root contact time decreases when the contact opening velocities are between 1 m/s and 10 m/s;

- The contact gap that the arc root moves from the contact region is less than indicated in [35];

- The arc root contact time can be made the same for high and low contact opening velocities by adjusted the gap behind the moving contact, and changing the contact and arc chamber materials.

8.3 Conclusion from experimental results

8.3.1 The gap behind the moving contact

A study of the influences of the gap behind the moving contact on the arc root motion using the AIS, FTA and pressure transducers was carried out. It was determined that the gap behind the moving contact causes a longer arc root contact time on the moving contact; this can be reduced by closed the gap. This considerably increases the energy transfer to the moving contact arc runner, suggesting an extension of the conductive arc column, including the establishment of a new electron receptor site.

Studies of the gap behind the moving contact on the arc root motion, flow velocity measurements from the pressure transducers and the thermal energy in the arc chamber lead to the following conclusion:

- The mass flow rate, the volume flow rate and flow velocity of the gas could be estimated;

- The arc root contact time was lower when the gap behind the moving contact was closed;

- The pressure behind the moving contact started to rise before the pressure behind the arc stack

8.3.2 The Short circuit current

The investigation of the influence of the short circuit current level on the arc root contact time has been studied. It was found that the short circuit current showed a significant influence on the movement of the arc root and the magnetic forces on the contact area. When the short circuit current increased, it induced an increase in the electromagnetic field. The self-blast magnetic field generated a high magnetic force.

The arc root contact time analysis and modelling of the magnetic forces lead to the following conclusions:

- The magnetic forces on the arc root increased as the arc current increased.

- The magnetic forces on the arc root were mainly dependent on arc current; the contact gap had a minimal effect.

- The short circuit current had a significant influence on the movement of the arc root from the contact region.

8.3.3 The contact opening velocity

The investigation of the influence of the contact opening velocity on the arc root contact time and pressure in the arc chamber has been described, with the magnetic forces dominating the arc root movement in the arc chamber. When the arc root moved from the contact region, the driven flows could dominate the velocity of the arc root.

The analysis of the arc root contact time, magnetic forces, arc velocity from the optical data and arc energy lead to the following conclusion:

- The contact opening velocity has a significant influence on the mobility of the arc root from the contact region.

- The arc current at the point at which the arc root moves from the contact region increased as the contact opening velocity was decreased.

- The arc root contact time decreased as the contact opening velocity was increased.

- Magnetic forces have a strong influence of the arc current, but a minimal effect on the contact gap.

- The high opening velocity has a high energy, high mass and high temperature rise.

8.3.4 The arc chamber venting

A study of the effects of the arc chamber venting on the arc root motion and the pressure has been carried out. The arc root motion depended upon the quantity of the flow and area of the arc chamber venting. This vent area of the arc chamber is critical to the arc root motion away from the contact region.

The analysis of the arc root contact time and the pressure in the arc chamber lead to the following conclusions:

- The influence of the arc chamber venting had an effect on the arc root of the moving contact.

- The arc root contact times increased as the vent area was decreased.

- The vent area of the arc chamber was critical to the arc root's movement away from the contact region

- The arc root motion depended upon the quantity of the flow and the area of the arc chamber venting.

8.4 General comments

8.4.1 Instrumentation and methodology

The optical fibre spectrometer was used to measure the spectral data during the arcing. It is recommended that this instrument should have an automatic trigger and variable integration time.

8.4.2 Experimental parameter

A study of the influence of the contact opening velocity on the arc root contact time, contact gap and arc current, at the time that the arc root moves from the contact region, should be repeated with more variation of the contact opening velocity. This would increase the accuracy of the minimum forces required for the arc root to move from the contact region.

8.4.3 Magnetic and gas dynamic forces modelling

The modelling of the magnetic forces and gas dynamics could be made more accurate by including the effects of current density, and the type and geometry of the contact and conductors.

REFERENCES

1. R.T. Lythall, "The switchgear book", Butterworth & Co., Ltd, London, 1972.

2. P.A. Jeffery," The motion of short circuit arcs in low current limiting miniature circuit breakers", Book submitted for PhD, University of Southampton, Jan. 1999.

3. Warren B Boast, "Vector fields", Harper & row, London, UK, 1964.

4. J.M.Somerville, "The Electric Arc", Methuen & Co., LTD, London, 1959.

5. Cassie A.M., "*Power System Switchgear-The Physical Nature of Arcs*", The electrical Journal 3, April 1959.

6. Paul G. Slade, "Electrical contacts Principles and Applications", Marcel Dekker, inc., New York, 1999.

7. F. Edlmayr, K. Krause, S. Theodor, " Low voltage switchgear", Heyden & Sons Ltd., London, 1973.

8. J.S. Morton, A*ir-breaker circuit breaker, Power Circuit breaker Theory and design*, Institution of Electrical Engineering, England, 1975, pp.185-187.

9. P. M. Weaver, J. W. McBride, "Arc motion in current limiting circuit breakers", 16 International conference on Electrical Contacts, Sep 1992, UK, pp.285-288.

10. J. W. McBride, " Investigations of arcing in a miniature circuit breaker under short circuit conditions, IC-ECAAA Conference Xian, China, May 1993, pp.419-423.

11. P. M. Weaver, J. W. McBride, " Magnetic and gas dynamics effects on arc motion in miniature circuit breakers", IEEE Holm Conference in electrical contact, Pittsburgh, Sep 1993, pp. 77-86.

12. P. M. Weaver, J. W. McBride," An optical fibre imaging system for the study of high speed motion", Sensors VI, Manchester, Sep. 1993, pp.227-232.

13. J.W. McBride, P.A. Jeffery, P.M. Weaver, " Electrode process and arc form in miniature circuit breakers",44 IEEE Holm Conference in Electrical contacts, Chicago, 1998, pp.93-99.

14. P.A. Jeffery, J.W.McBride, " Anode and cathode arc root movement during contact opening at high current", IEEE Transactions on components packaging and manufacturing technology ,part A, Dec. 1998.

15. E.M.Belbel, M.Lauraire, "Behaviour of switching arc in low voltage limiter circuit breakers", IEEE Trans. Comp. Hybrids, Manufact, Technol, Vol. CHMT-8, pp.3-12, 1985.

16. M. Abbaoui, B. cheminat, " Determination of the charaterstics of an electric arc plasma contaminated by vapors from insulators", IEEE Transaction on Plasma Science, Vol. 19, no. 1, Feb. 1991, pp. 1-8.

17.John J. "Dielectric recovery characteristics of a high current arcing gap", 47th IEEE Holm Conference in Electrical contacts, Montreal, Canada, 2001, pp.154-160.

18.John Shea, " The influence of arc chamber wall material on arc gap electric recovery voltage", IEEE2000,.

19.G. Velleaud, A. Laurent, M. Mercier, " a study of the kinetics of a low voltage breaking self blown arc: Analysis of the derivative of the anode cathode voltage", J. Phys. D: Appl, Phys. 22, 1989, pp. 933-940.

20.G. Velleaud, M. Mercier, A. Laurent, F. Gary, " Use of an inverse method on the determination of the evolution of a self blown electric arc in the air", IEEE Transactions on plasma science, vol. 19, No. 3, June 1991, pp.510-514.

21.G.Meunier, A. Abri, " Simulation of the arc interruption in circuit breaker", The 5th Switching arc phenomena, Poland, Sept. 1985, pp. 105-109.

22.F. Karetta, M. Lindmayer, " Simulation of the gas dynamic and electromagnetic processes in low voltage switching arcs", 42nd IEEE Holm Conference in Electrical contacts, Chicago, 1996, pp.35-45.

23.E. Belbel, L. Siffroi, "Immobility duration of electric arcs between contactor poles at breaking instant", Electrical contact 1982,pp. 168-170.

24.Rieder W., " Low current arc modes of short length and time", IEEE Transaction On Components Packaging and Manufacturing Technology, Vol.23, No. 2, June 2000, pp. 286-292.

25. P.A. Jeffery, J.W. McBride, "*Electrode Processes and Arc Form in Miniature Circuit Breakers*", 44th IEEE Holm Conference on Electrical Contacts, Chicago 1998, pp 93-99.

26. J.W. McBride, P.W. Weaver, P.A. Jeffery, "*Arc Root Mobility During Contact Opening at High Current*", IEEE Transaction On Components Packaging and Manufacturing Technology Part A, Vol.121, No. 1 March 1998, pp. 61-67.

27. J.W.McBride, "High speed, medium resolution arc imaging in current limiting devices", 17 International Conference on Electrical Contacts, Japan, July, 1994.

28. J. M. McBride, P.A. Jeffery, "The design optimisation of current limiting circuit breakers", Proc. 3 Int. Elect. Cont. Arcs., Apparatus and Applications, ECAAA, China, 1997, pp.354-360.

29. J.W.McBride, P.M.Weaver, "Arc root mobility during contact opening at high current", IEEE Transactions on Components Packaging and Manufacturing Tyechnology, Part A, Vol. 21, No.1, March 1998, pp61-67.

30. P.A. Jeffery, J.W. McBride, J. Swingler, P.M. Weaver, " An investigation into arc contact immobility and current limiting performance of miniature circuit breakers using the Taguchi design of experiments", 19 International conference on Electrical contact Phenomena, ICEC98, Germany, Sep. 1998.

31. P.A. Jeffery, J.K.Sykulski, J.W. McBride, "3D Finite element analysis modelling of the arc chamber of a current limiting miniature circuit breaker", Compel-The international journal for computation and

mathematics in electrical and electronic engineering, vol.17, no.1/2/3, 1998, pp.224-251.

32. J.W.McBride, P.M.Weaver, P. Jeffery, "Arc root mobility during contact opening at high current", Proc. 18 Int. Conf. Electr. Cont. Phen, 1996, pp.27-34.

33. Poeffel K.,"Influence of the copper electrode surface on initial arc movement", IEEE. Trans. PS 8(1980), pp 43-448.

34. McBride J.W, Jeffery P.A, "Anode and Cathode Arc Root Movement during Contact Opening at High Current", IEEE Transactions on Components Packaging and Manufacturing Technology, Vol. 22, No 1, March 1999, pp. 38-46. Additional supplement, McBride J.W, Jeffery P.A, "Anode and Cathode Arc Root Movement during Contact Opening at High Current", IEEE Transactions on Components Packaging and Manufacturing Technology, Vol. 22, No 2, pp.344, June 1999.

35. Rieder W., "Interaction between magnet blast arcs and contacts ", Electrical contact, 1982, 28th Holm Conference, Illinois, Sep. 13-15, pp. 3-10.

36. Gauster E, Rieder W, "Arc wall interaction phenomena immediatedly after contact separation in magnet-blast interrupters", Proc. 41 IEEE Holm Conf.(1995), pp.365-372.

37. Rieder W, Veit C, Gauster E," Influence of materials on the interaction of switching arcs with contacts and walls, IEEE Trans. CPMT 15,(1992), pp.1123-1137.

38. Limdmayer M., "The influence of Contact Materials and Chamber wall material on the migration and the splitting of the Arc in extinction chambers", IEEE Trans Hybrids and Packing, Vol. PHP-9, No. 1, March 1973.

39. Widmann W. "Arc commutation across a step or a gap in one of two parallel copper electrodes", *IEEE Trans. Components, Hybrids and Manufacturing Technology,* Vol. CHMT-8, No1, March 1985, pp.21-28.

40. M. Lindmayer, M. Springstubbe, " 3D-simulation of arc motion between arc runners including the influence of ferromagnetic material, 47 IEEE Holm Conf. on Electrical Contacts, 2001, Montreal, pp. 148-153.

41. D. Podolsky, V. Kapustin, " Study of electric field between two parallel electrodes with arc at a low voltage", 44 IEEE Holm Conf. on Electrical Contacts, 1998, New York, pp. 301-306.

42. Gauster E., Rieder W., "Arc Lengthening between divergent runners: influence of geometry and materials of runners and walls", 42nd IEEE Holm Conf. on Electrical Contacts, 1996, pp. 1-11.

43. Fievet C., Siriex t., Andre V., Fleurier C.,"*Optical Diagnostics of Arc Transfer Phenomena in Low Voltage Circuit Breakers*", 3rd International Conference on Electrical Contacts, Arc Apparatus and their Applications (IC-ECAAA), Xi'an, China, May 19-22, 1997.

44. P. Kovita, " Ablation-Stabilized arc in Nylon and Boric Acid Tubes", IEEE Transaction on plasma science, Vol. PS-15, No.3, June 1987, pp. 294-301.

45. Yutaka Takenaka, Kenji Funaki, Shingo Shimada," Effects of organic Gas Components on Contact Resistance", Proc. IEEE 1995,pp.260-265.

46. Ozlem Mutaf-Yardimci, Alexei V. Savelier, Alexander A. Fridman, Lawrence A.Kennedy, " Thermal and nonthermal regiomes of gliding are discharge in air flow", Journal of Applied Physics, Vol. 87, No. 4, 15 Feb. 2000, pp. 1632-1641.

47. V. Behrens, Th. Honig, A. Kraus, E. Mahle, R. Michal, K.E. Saeger, " Test results of Different Silver/Graphit contact materials in Regard to Applications in Circuit breakers", Proc. IEEE 1995, pp. 393-397.

48. Philip C Wingert, " The effects of interrupting elevated currents on the erosion and structure of silver-graphite", Proc. IEEE 1996, pp. 60-69.

49. Hirose K, "Immobility phenomena of the DC electric arc of large current driven by magnetic field", Electrotechn. J. japan 7(1962), pp.58-64.

50. Martin Bizjak, "Influence of vapour pressure on the dynamics of repulsion by contact blow off", Interconf. 2002, pp.268-275.

51. Kiyoshi Yoshida, Atsuo Takahashi, "Spectroscopic Measurement of Ag Breaks Arc", in Proc. Int. Conf. on Electrical Contact, Japan, 1994, pp. 51-58.

52. M. Takeuchi, T. Kubono, " The spatial distribution of temperature in a cross section column nearby cathode contact", IEEE 1995, pp.210-218.

53. K. Sato, T. Sato, T. Takagi, " Deveolpment of a high speed time resolved spectroscope and its application to analsis of time varying optical spectra", IEEE. Transaction on Instrument and measurement, Vol. 36, No.4, Dec. 1987, pp. 1045-1049.

54. C. Fievet, L. Sirieix, V. Andre, C. Fleurier, "Optical Diagnostics of arc transfer phenomena in low voltage circuit breakers", 3 Int . conf. on Electrical contact, arcs, Apparatus and their application IC-ECAAA, May 1997, pp. 19-22.

55. G. Velleaud, M. Mercier, A. Laurent, F. Gary, " Use of an inverse method on the determination of the evolution of a self blown electric arc in the air", IEEE Transactions on plasma science, vol. 19, No. 3, June 1991, pp.510-514.

56. G. Velleaud, A. Laurent, M. Mercier, " a study of the kinetics of a low voltage breaking self blown arc: Analysis of the derivative of the anode cathode voltage", J. Phys. D: Appl, Phys. 22, 1989, pp. 933-940.

57. Ragnar Holm, "Electric contact, 4Edition, Spring Verlag, 1981, pp.316-337.

58. S. Lei, S. Garimella, S. Chan, " Gas dynamic and electromagnetic processes in high current arc plasmas", Journal of Applied Physics, Vol.85, No. 5, March 1999, pp. 2547-2555.

59. P. Boddy, T. Utsumi, " Fluctuation of arc potential caused by metal-vapor diffusion in arcs in air", Journal of Applied Physics, Vol. 42, No.9, Aug. 1971, pp.3369-3373.

60. C. Fievet, P.Petit, M. Perrin, P. chevrier, G. Bernard, "Residual conduction in low voltage circuit breakers", 11 Internation conference on gas discharges and their applications, " Tokyo, Sep, 1995, pp. 26-29.

61. J. Shea, D. Boles, Y. Chien, R. Zeigler, " Computer animated digital arc diagnostic system", IEEE Transaction on components packaging and manufacturing technology, Part A, Vol. 17, No. 1, march 1994, pp. 47-52.

62. P.Kovitya, "Physical properties of high pressure plasmas of hydrogen and copper in the temperature range 5000-60000K, IEEE Transaction on Plasma Science, Vol. PS13, No. 6, Dec. 1985, pp. 587-594.

63. C.B. Ruchti, L. Nimeyer, "Ablation controlled arcs", IEEE Transactions on Plasma Science, Vol. PS14, No. 4, Aug. 1986, pp.423-434.

64. Yardimci, A. Saveliev, A. Fridman, L. Kennedy, "Thermal and nonthermal regimes of gliding arc discharge in air flow", Journal of applied Physics, Vol. 87, No. 4, Jan. 2000.

65. M. Lindmayer, J. Paulke, " Arc motion and pressure formation in low voltage switchgear", IEEE 1996, pp. 17-26.

66. McBride J.W, Weaver P.M, "Review of Arcing Phenomena in Low Voltage Current Limiting circuit Breakers", IEE Proc.-Sci. Meas. Technol., Vol. 148, No. 1, January 2001, pp.1-7.

67. Operating manual and user's guide, S2000 miniature fiber optic spectrometers and accessories. WWW.OceanOptics.com.

68.B. Ravindranath, *Power System Protection and Switchgear*, Wiley eastern limited, India, 1977, pp 284-298.

69.WWW.chemicalelements.com

70.J.G.J. Sloot and G.M.V.Bosch, "Some conditions for arc movement under the influence of a transverse magnetic field", Holectechniek, 1972, 98-105.

71.Selected constant Metals thermal and mechanical data, May, 1961.

72.P.Kirchesch, L.Niemeyer, "Arc behaviour in an ablating nozzle", 5 int. symp. Switching arc Phenomena, Lodz, Poland, Sep. 1985, pp.24-26.

73.C.B. Ruchti, L. Nimeyer, "Ablation controlled arcs", IEEE Transactions on Plasma Science, Vol. PS14, No. 4, Aug. 1986, pp.423-434.

74.N. Behrens," Arc motion between opening and diverging electrodes", Proc. of Electrical contact phenomena Conf., 1978, pp. 243-247.

75.A.E. Guile, T.J.Lewis, P.E.Secker," The motion of cold cathode arcs in magnetic fields", Proc. IEE part C 108, 1961, pp.463-470.

76.A.E.Guile, S.F.Mehta,"Arc movement due to the magnetic field of current flowing in the electrodes", IEE, paper no. 2413, Dec. 1957, pp. 533-540.

77.J.W.McBride, S.M.A.Sharkh, "The effect of contact opening velocity and the moment of contact opening on the AC erosion of Ag/CdO contacts", 39th IEEE Holm Conf. on Electrical Contacts, Pittsburgh, Sep. 1993, pp. 87-95.

78. Z.K.Chen, K.Sawa, "Effect of Arc behaviour on Material Transfer, IEEE Transaction on components packaging and manufacturing technology, Part A, Vol. 21, No. 2, march 1998, pp. 310-322.

79. W. Merck, V. Zatelepin, "The gas dynamics of current limiting devices during immobility time", IEEE Transactions on Plasma Science, Vo. 25, No. 5, Oct. 1997, pp. 947-953.

APPENDIX

This appendix shows the experimental results of the arc current, arc voltage, arc root displacement and arc root contact time as the influence of :

1) Contact configuration

   - Ag/C on the moving contact

   - Ag/C on the fixed contact

2) Supply polarity

   - Anode on the moving contact

   - Anode on the fixed contact

3) Arc chamber venting

   - Anode on the moving contact

   - Anode on the fixed contact

4) Contact opening velocity

   - 1 m/s

   - 4 m/s

   - 5.5 m/s

   - 10 m/s

5) Arc chamber material

- Polycarbonate

- Ceramic

6) Contact materials

- Ag/C

- Cu

A.1 Contact configurations: Ag/C on the moving contact

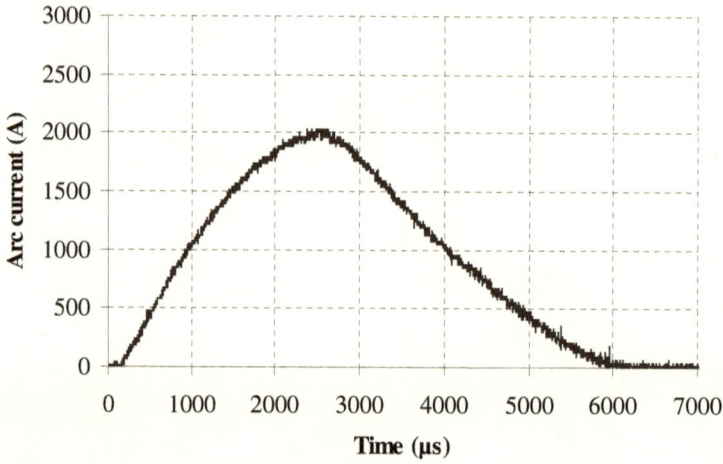

(A) arc chamber venting : Opened

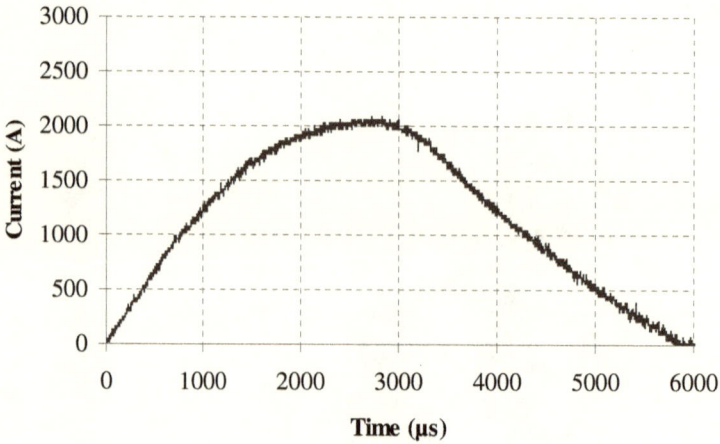

(B) Arc chamber venting : Choked

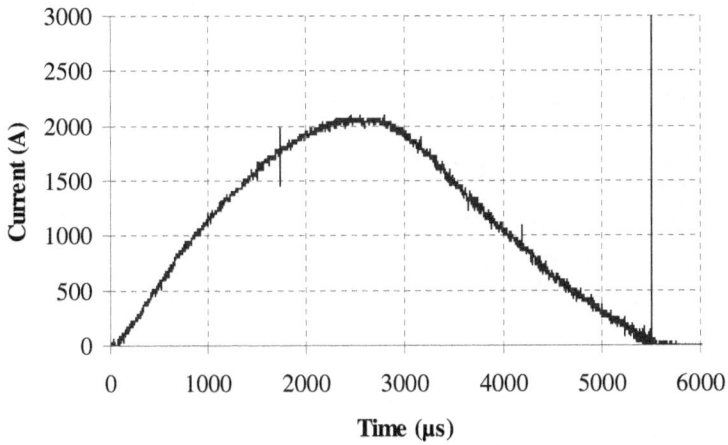

(C) Arc chamber venting : Closed

Figure A.1: Arc current waveform when Ag/C on the moving contact with opened, choked and closed arc chamber venting

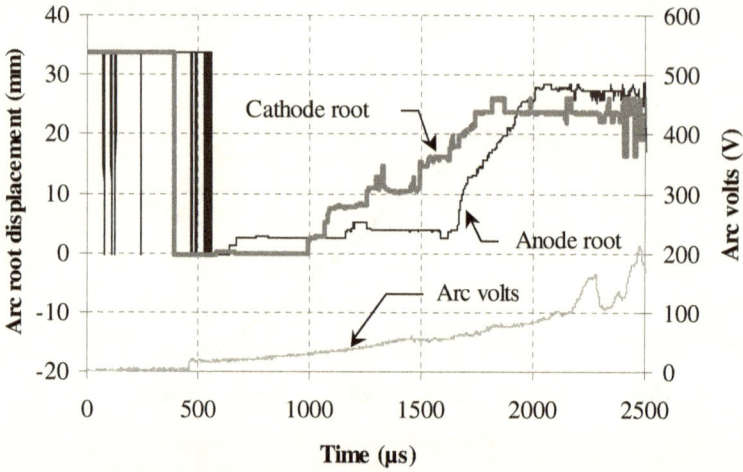

(A) Arc chamber venting : Opened

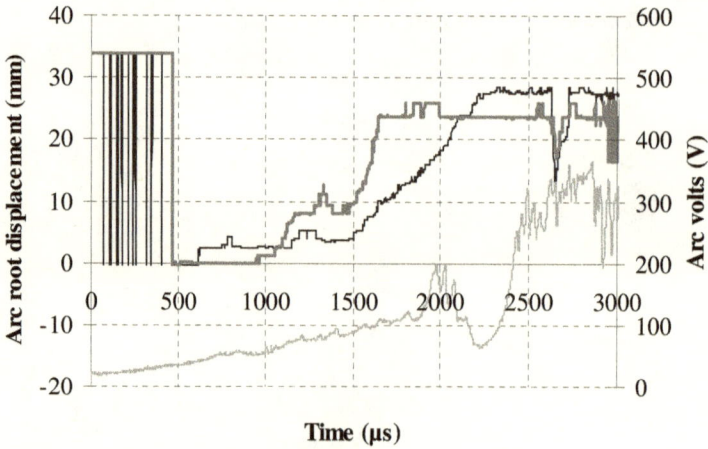

(B) arc chamber venting : Choked

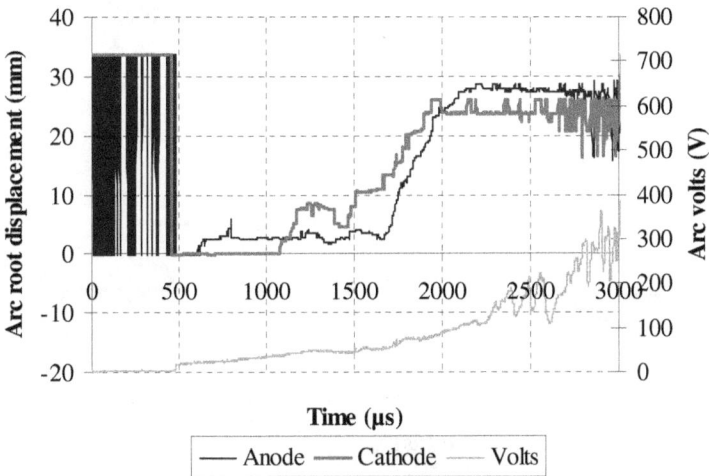

(C) Arc chamber venting : Closed

Figure A.2: Arc root displacement when Ag/C on the moving contact with opened, choked and closed arc chamber venting

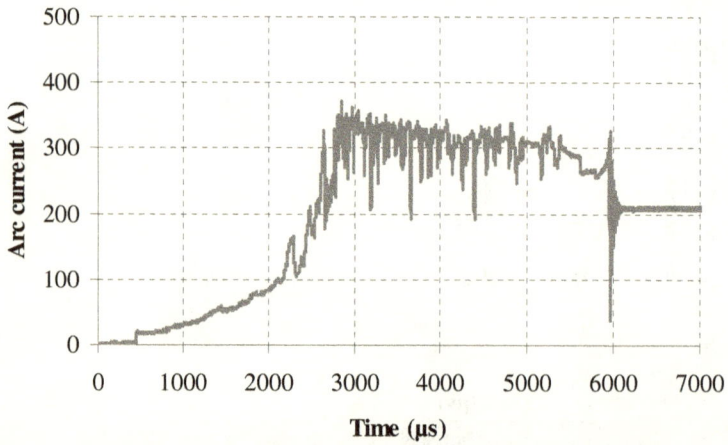

(A) Arc chamber venting : Opened

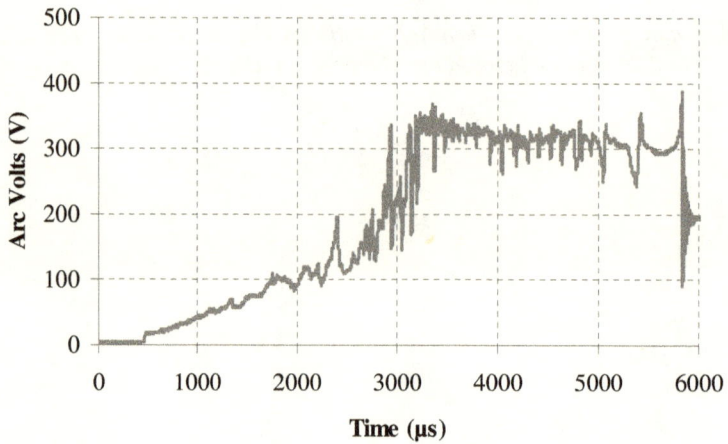

(B) Arc chamber venting : Choked

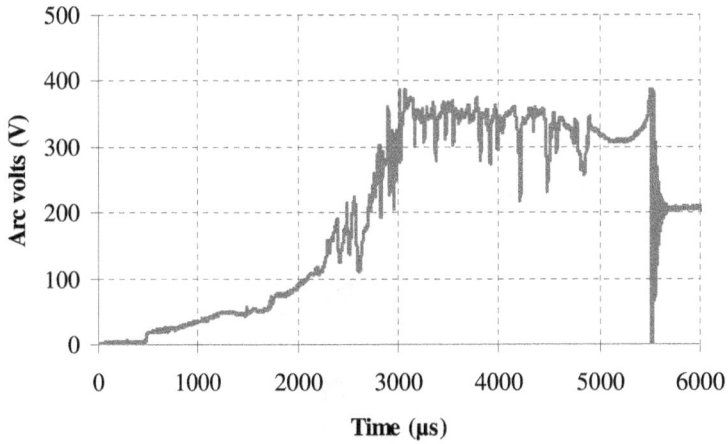

(C) Arc chamber venting : Closed

Figure A.3: Arc voltage waveform when Ag/C on the moving contact with opened, choked and closed arc chamber venting

A.2 Contact configurations: Ag/C on the fixed contact

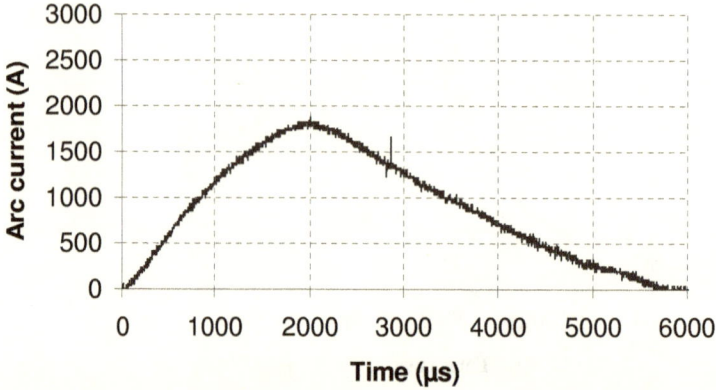

(A) Arc chamber venting : Opened

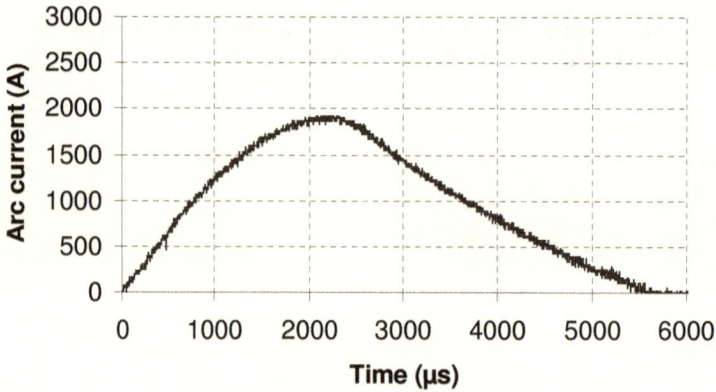

(B) Arc chamber venting : Choked

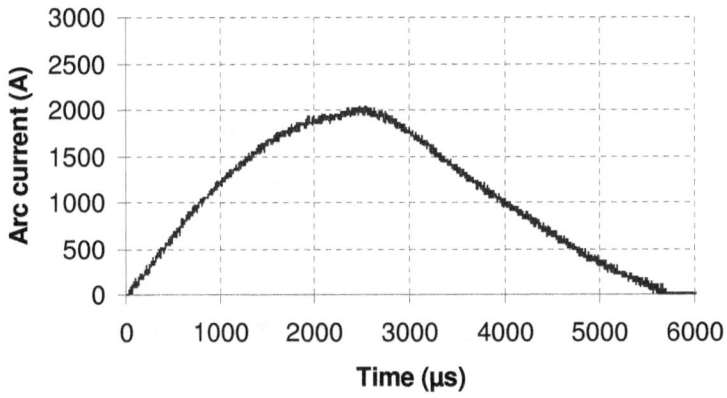

(C) Arc chamber venting : Closed

Figure A.4: Arc current waveform when Ag/C on the fixed contact

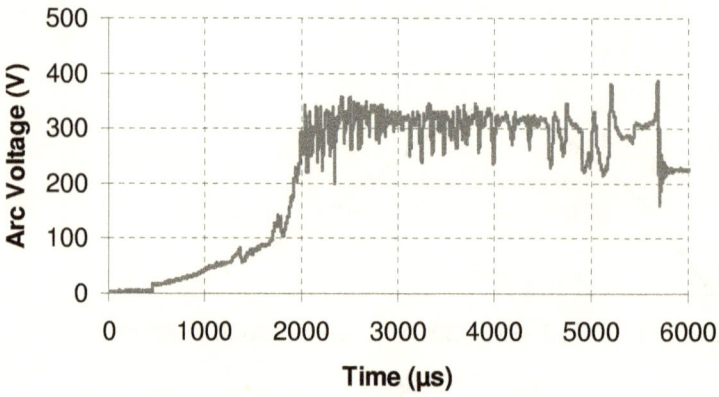

(A) Arc chamber venting : Opened

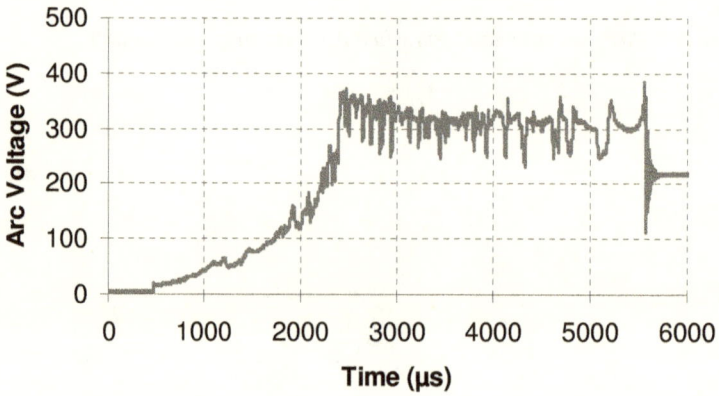

(B) Arc chamber venting : Choked

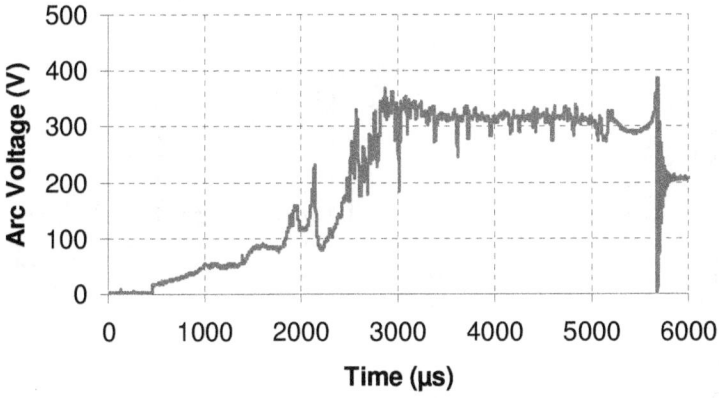

(C) Arc chamber venting : Closed

Figure A.5: Arc voltage when Ag/C on the fixed contact

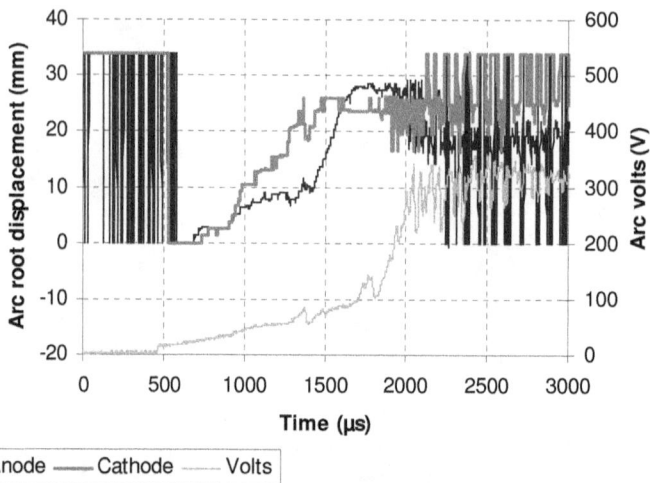

(A) Arc chamber venting : Opened

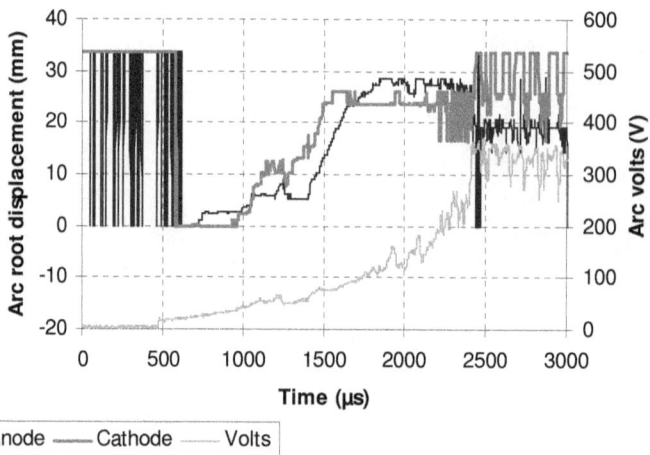

(B) Arc chamber venting : Choked

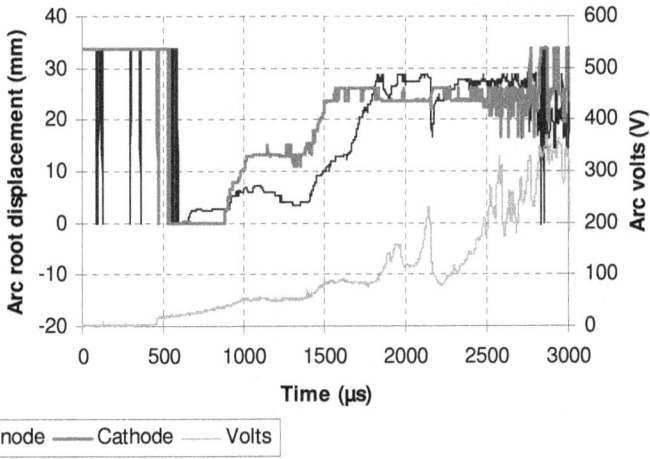

(C) Arc chamber venting : Closed

Figure A.6: Arc root displacement when Ag/C on the fixed contact

A.3 Supply polarity: Anode on the fixed contact

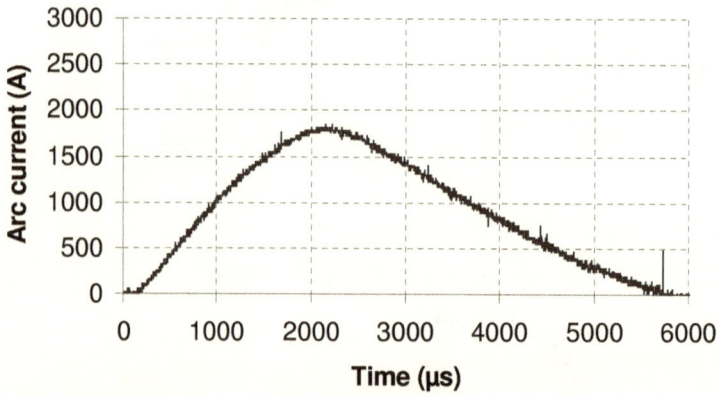

(A) Arc chamber venting : Opened

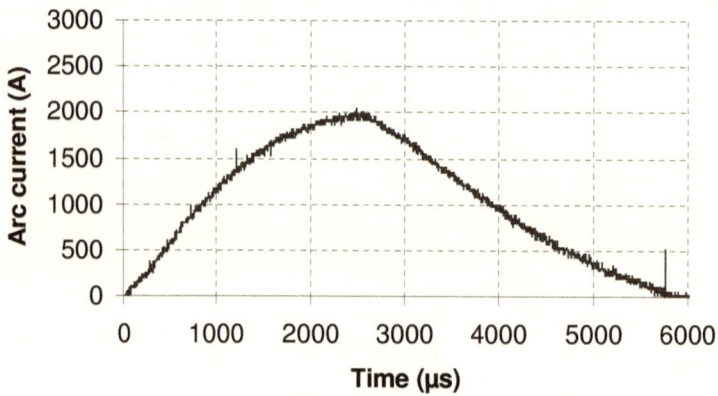

(B) Arc chamber venting : Choked

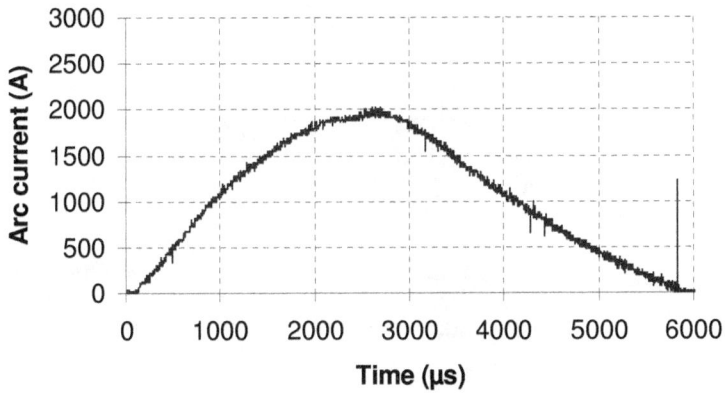

(C) Arc chamber venting : Closed

Figure A.7: Arc current waveform when Ag/C on the fixed contact, power supply anode on the fixed contact and opened, choked and closed arc chamber venting

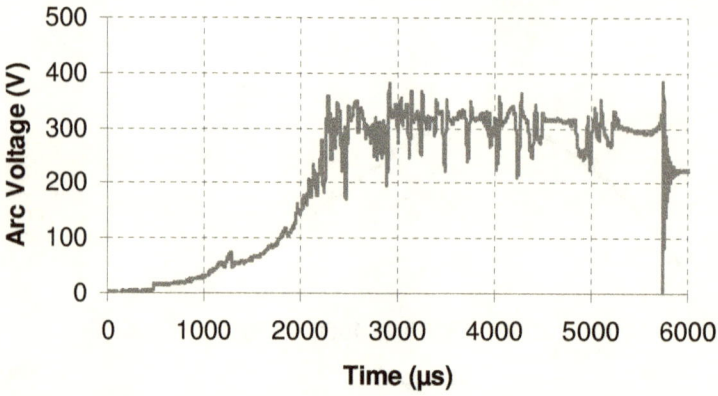

(A) Arc chamber venting : Opened

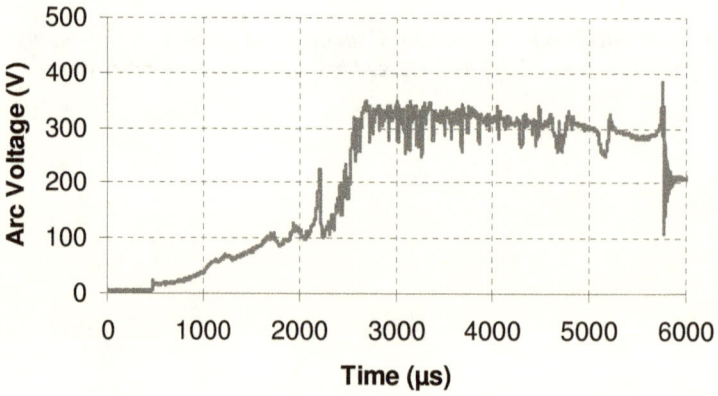

(B) Arc chamber venting : Choked

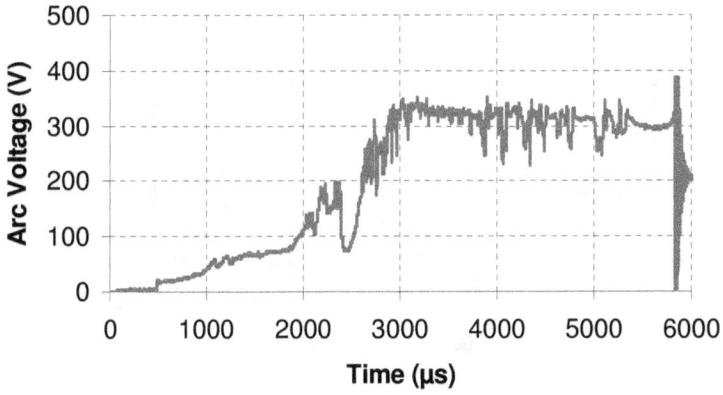

(C) Arc chamber venting : Closed

Figure A.8: Arc voltage waveform when Ag/C on the fixed contact, anode on the fixed contact and opened, choked and closed arc chamber venting

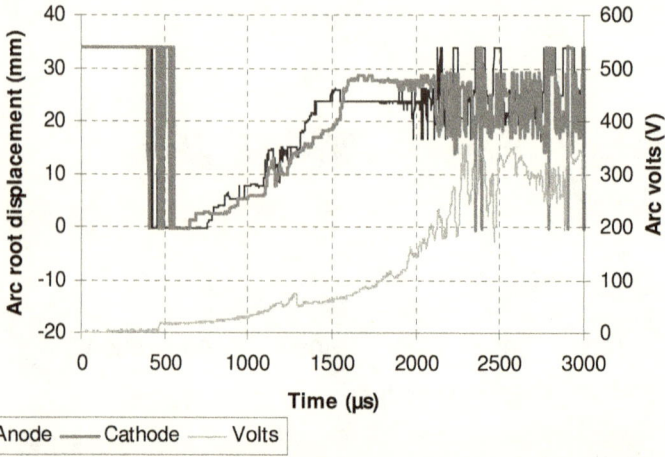

(A)Arc chamber venting : Opened

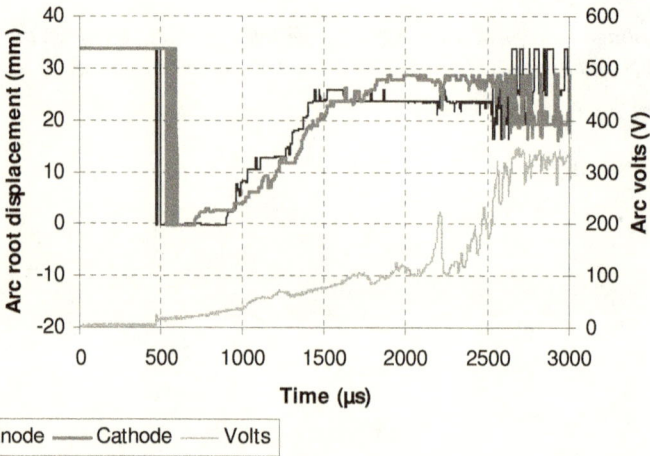

(B) Arc chamber venting : Choked

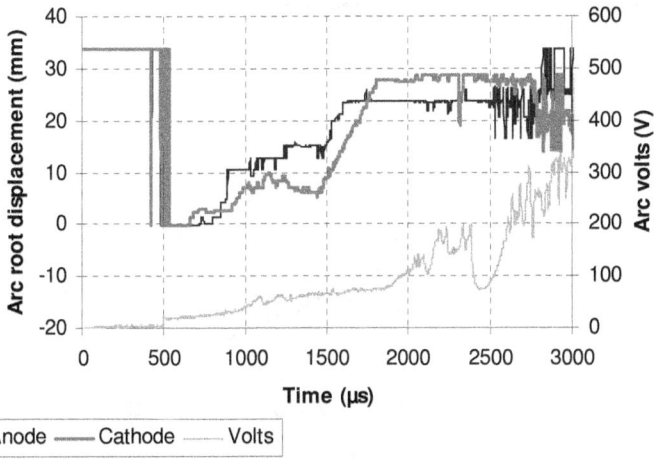

(C) Arc chamber venting : Closed

Figure A.9: Arc root displacement when anode on the fixed contact and opened, choked and closed arc chamber venting

A.4 Supply polarity: Anode on the moving contact

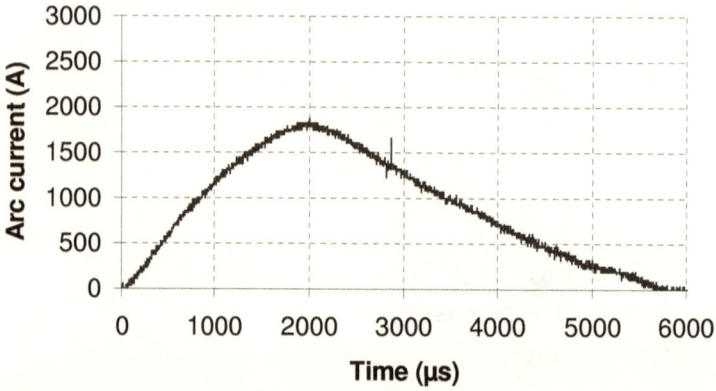

(A) Arc chamber venting : Opened

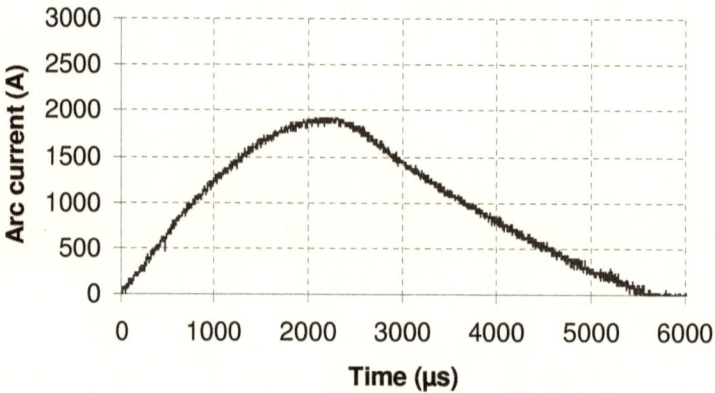

(B) Arc chamber venting : Choked

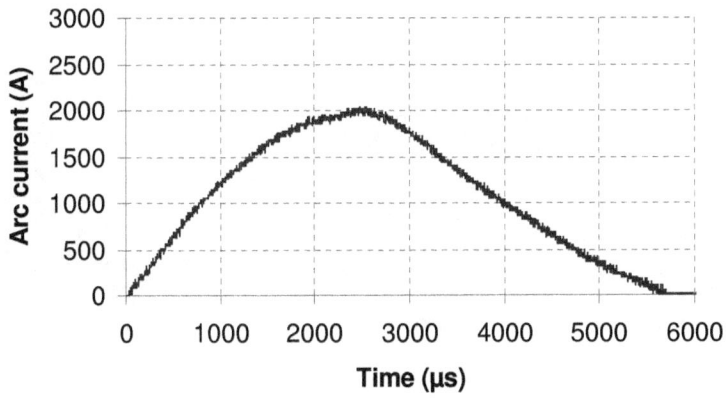

(C) Arc chamber venting : Closed

Figure A.10: Arc current when the anode power supply on the moving contact and opened, choked and closed arc chamber venting

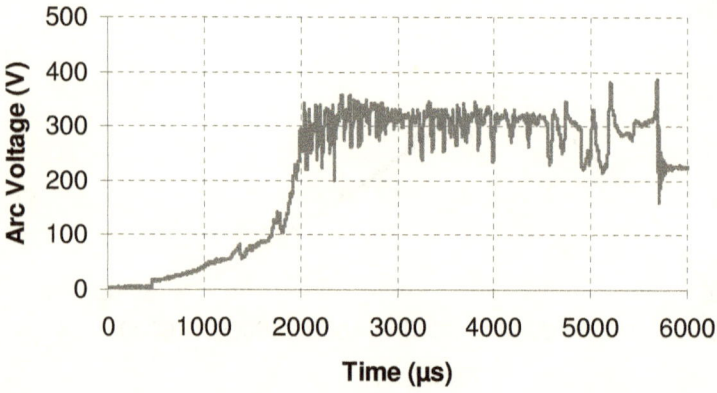

(A) Arc chamber venting : Opened

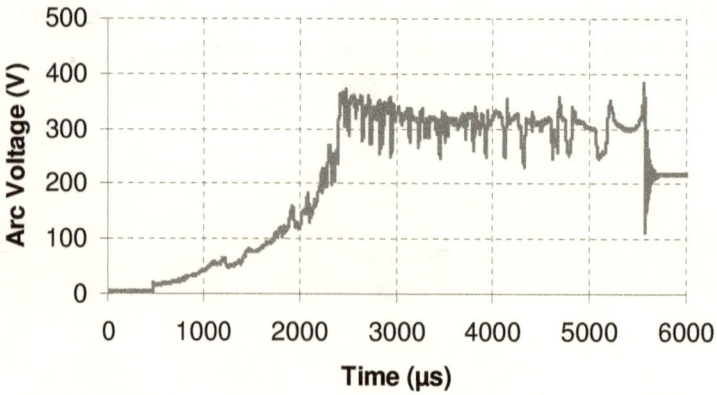

(B) Arc chamber venting : Choked

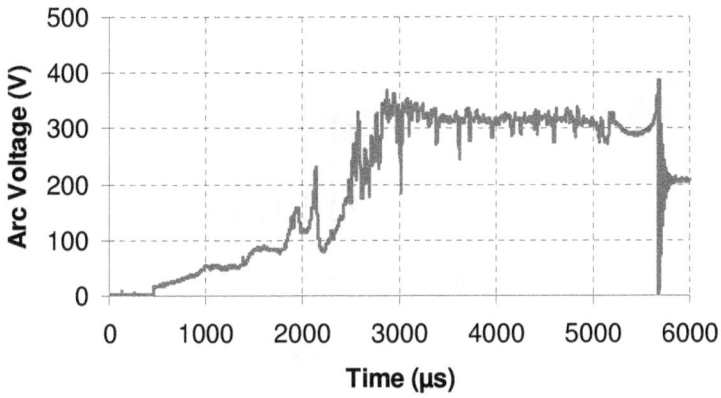

(C) Arc chamber venting : Closed

Figure A.11: Arc voltage when the anode power supply on the moving contact and opened, choked and closed arc chamber venting

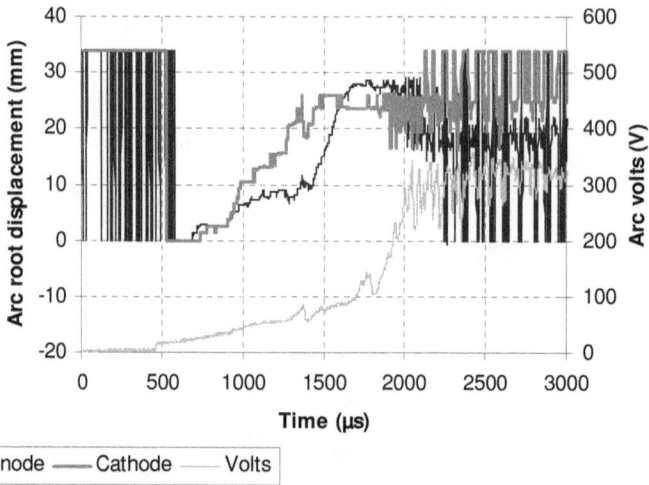

(A) Arc chamber venting : Opened

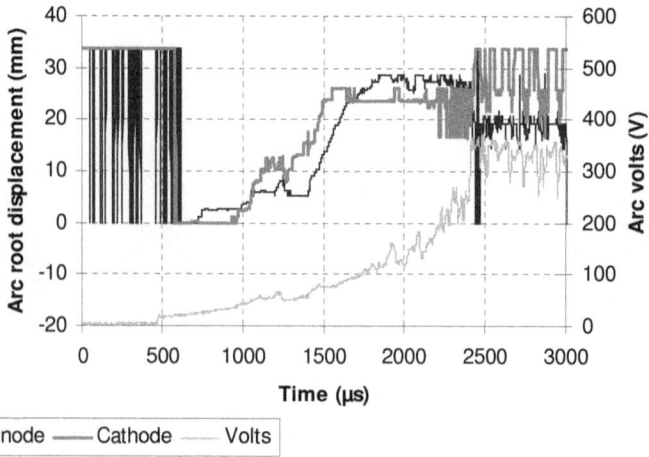

(B) Arc chamber venting : Choked

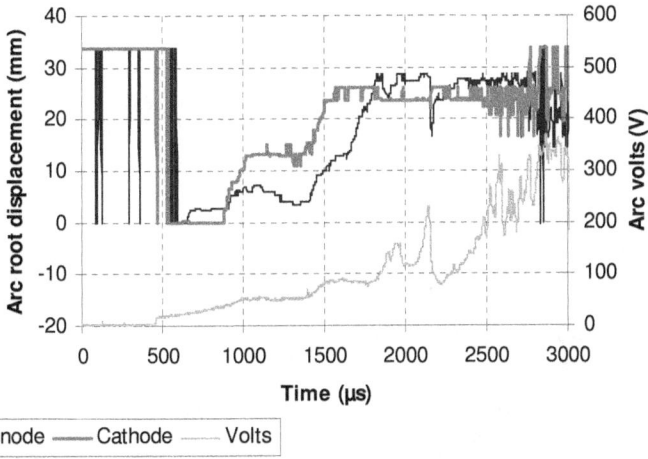

(C) Arc chamber venting : Closed

*Figure A.12: Arc root displacement when the anode power supply on the moving
contact and opened, choked, closed arc chamber venting*

A.5 Short circuit current level: Anode on the moving contact

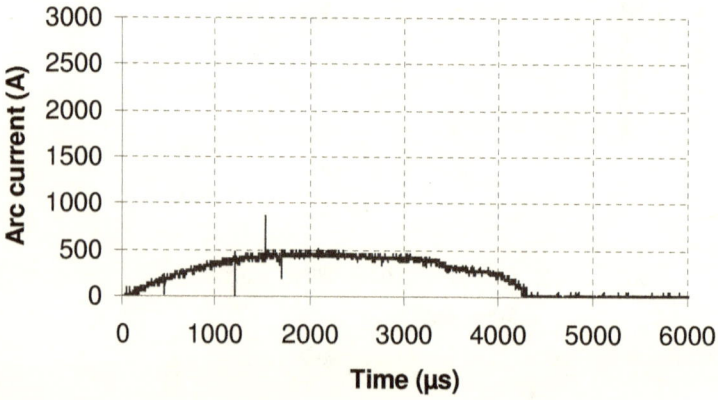

(A) Short circuit current : 500 A.

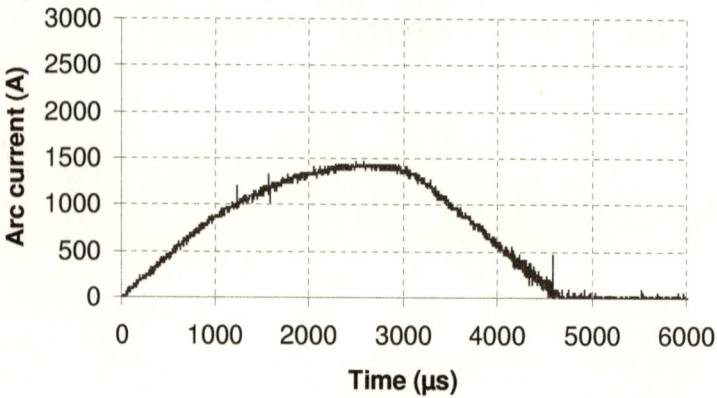

(B) Short circuit current : 1400 A.

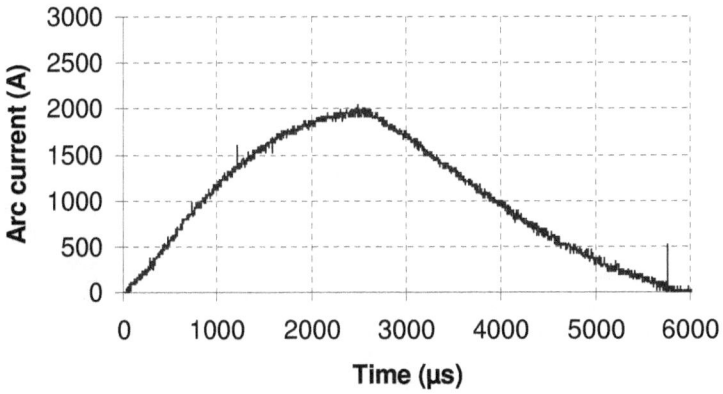

(C) Short circuit current : 2000 A.

Figure A.13: Arc current when arc current level 500A,1400A and 2000A

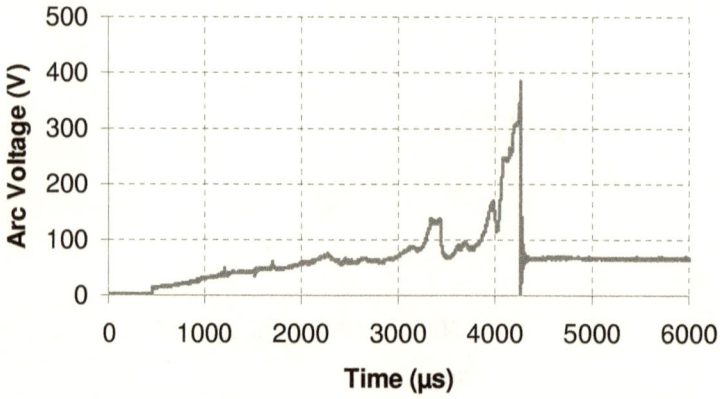

(A) Short circuit current : 500 A.

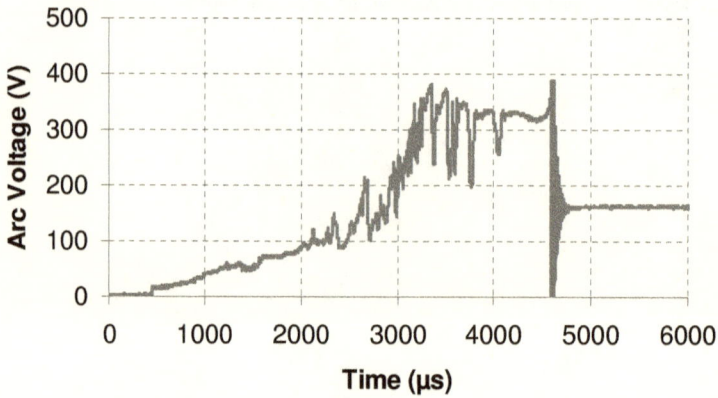

(B) Short circuit current : 1400 A.

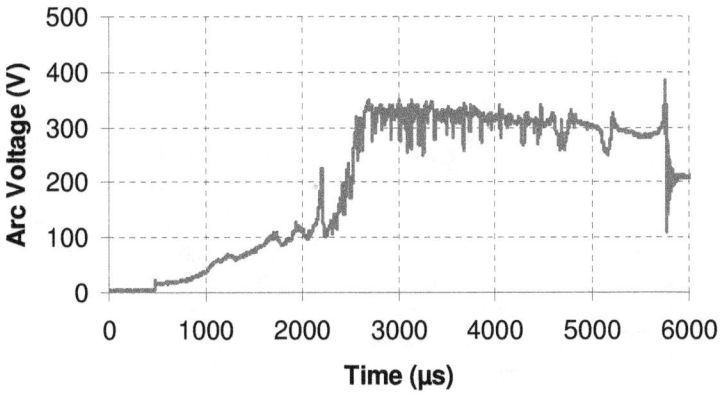

(C) Short circuit current : 2000 A.

Figure A.14: Arc voltage when arc current level 500A,1400A and 2000A

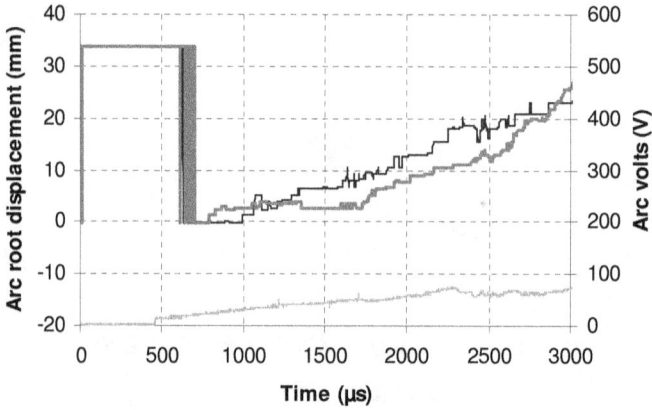

(A) Short circuit current : 500 A.

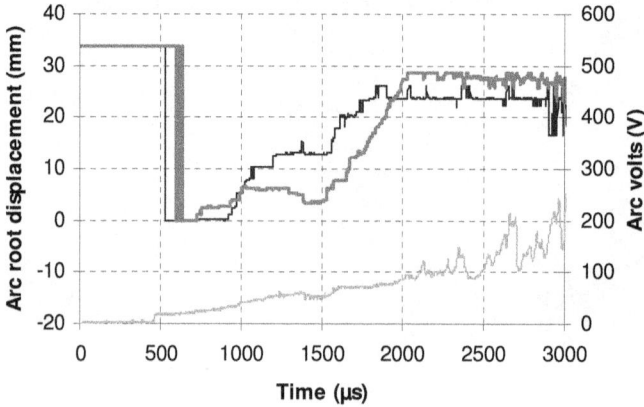

(B) Short circuit current : 1400 A.

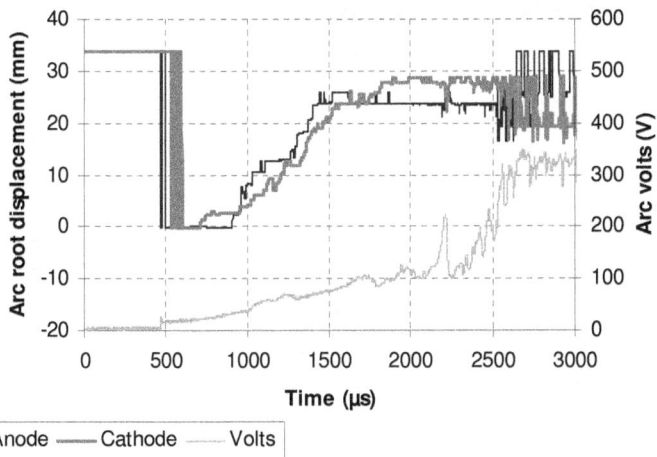

(C) Short circuit current : 2000 A.

Figure A.15: Arc root displacement when arc current level 500A, 1400A and 2000A

A.6 Arc chamber venting: Anode on the fixed contact

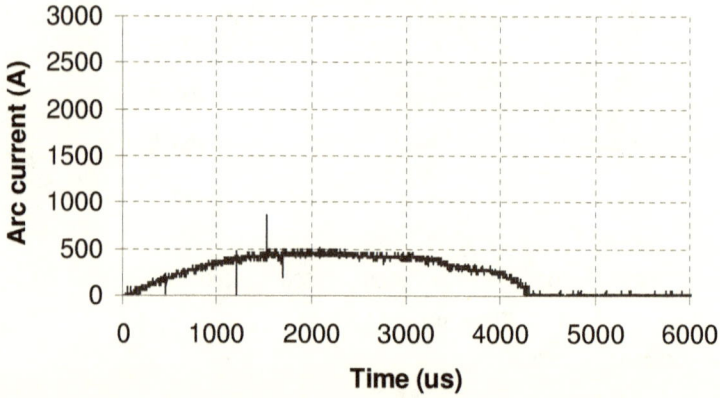

(A) Short circuit current : 500 A.

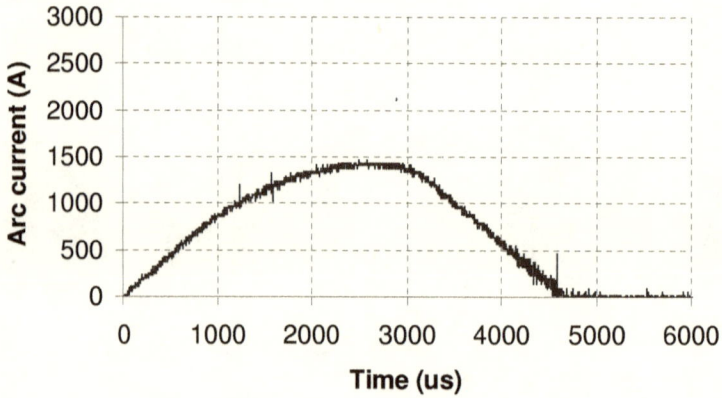

(B) Short circuit current : 1400 A.

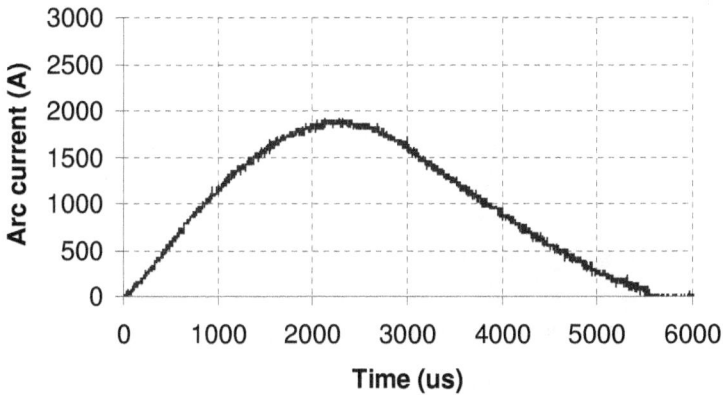

(C) Short circuit current : 2000 A.

Figure A.16: Arc current when short circuit current 500A, 1400A and 2000A

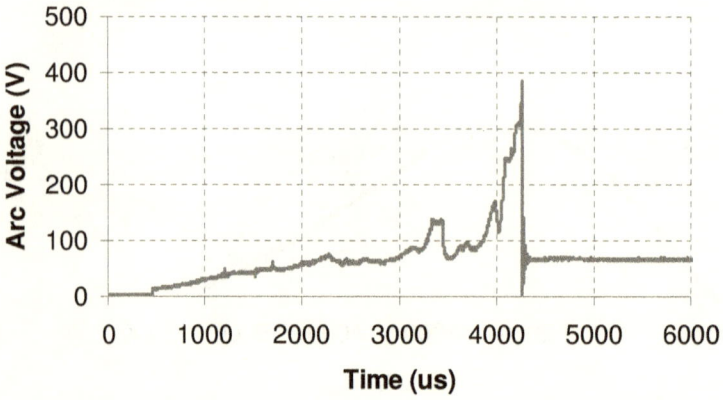

(A) Short circuit current : 500 A.

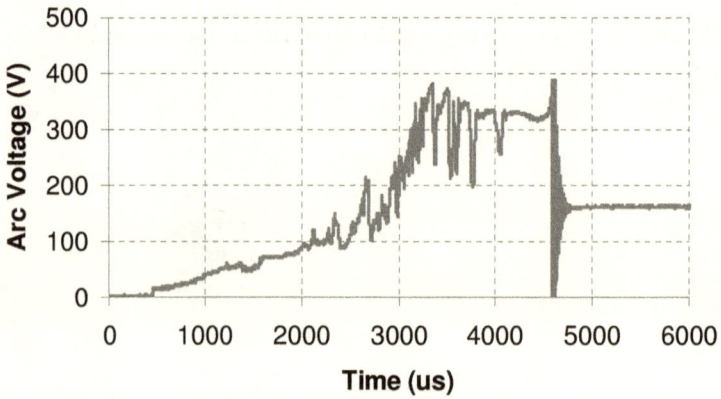

(C) Short circuit current : 1400 A.

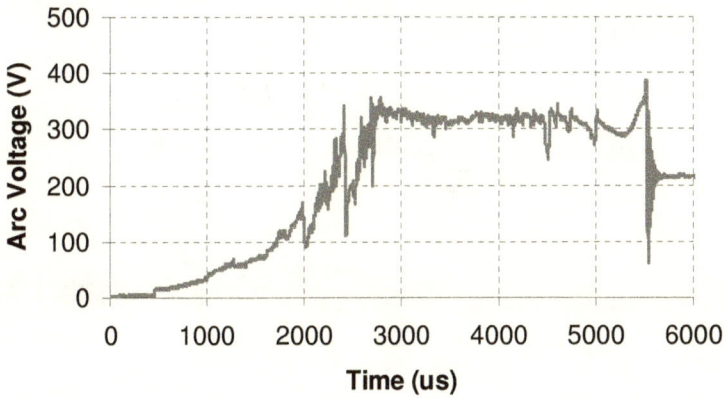

(C) Short circuit current : 2000 A.

Figure A.17: Arc voltage when short circuit current 500A, 1400A and 2000A

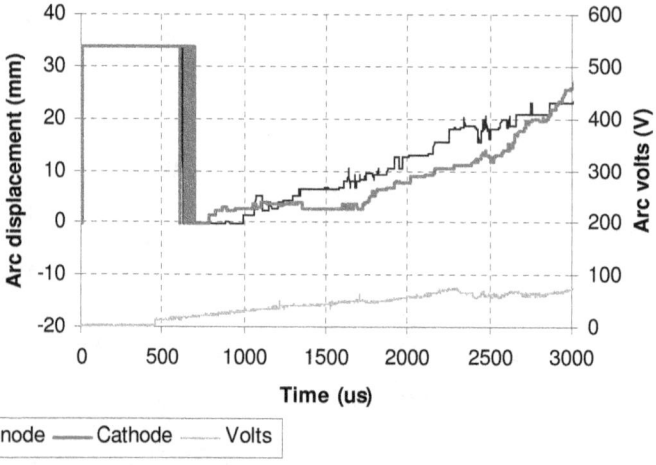

(A) Short circuit current : 500 A.

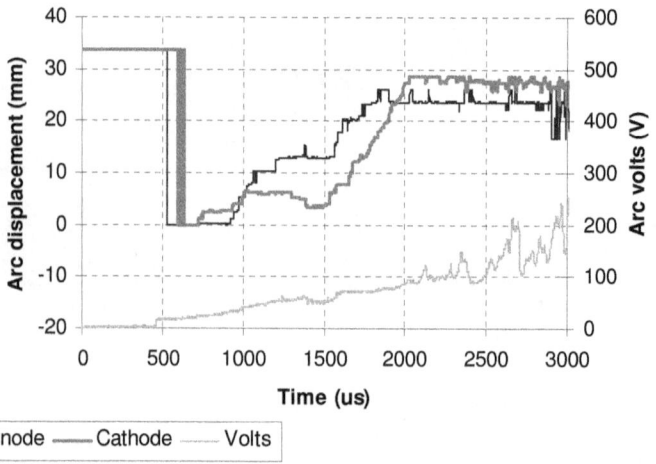

(B) Short circuit current : 1400 A.

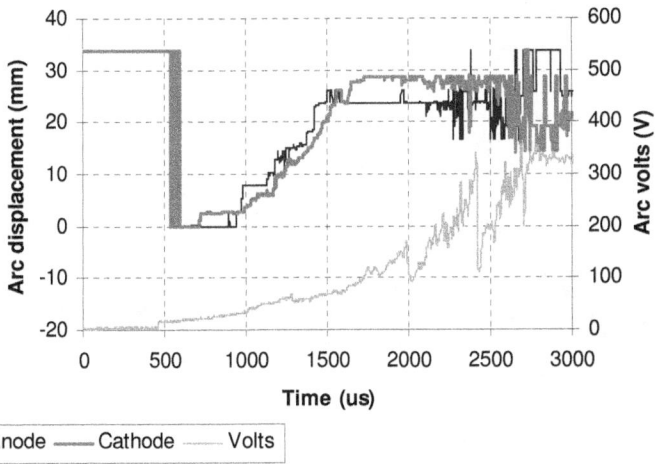

(C) Short circuit current : 2000 A.

Figure A.18: Arc root displacement when short circuit current 500A, 1400A, 2000 A

A.7 Contact opening velocity

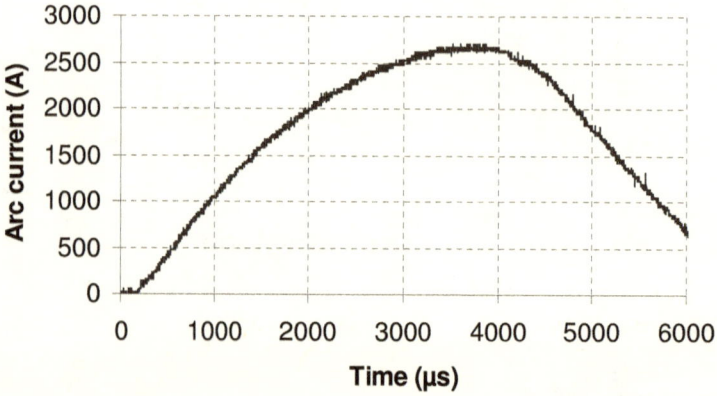

(A) Contact opening velocity : 1 m/s.

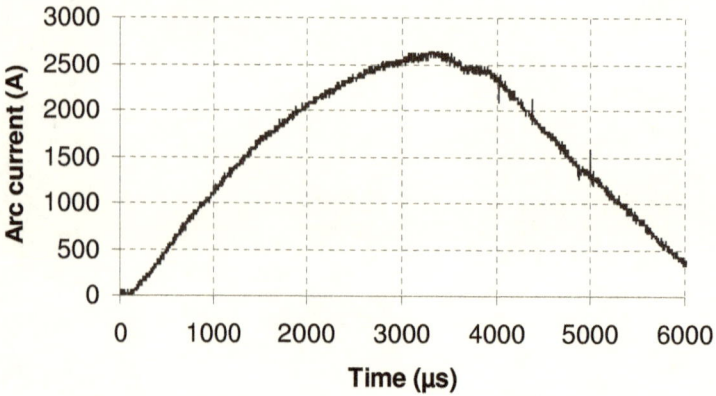

(B) Contact opening velocity : 4 m/s

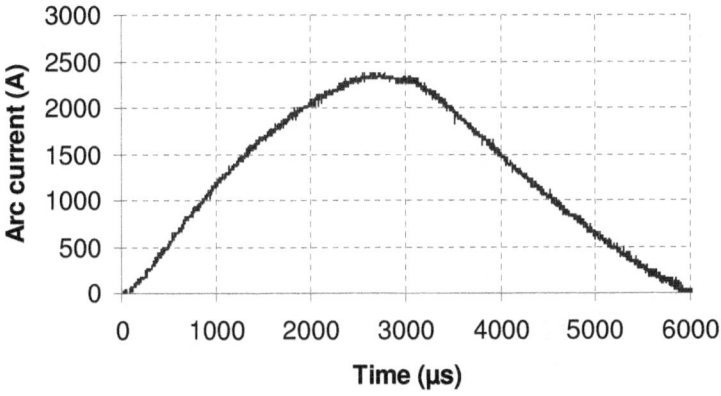

(C) Contact opening velocity : 5.5 m/s

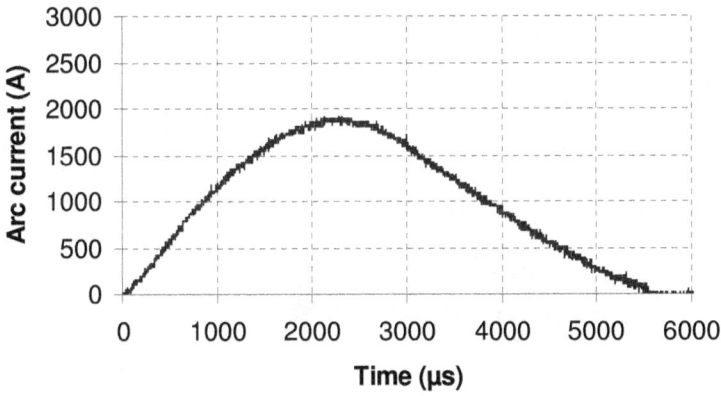

(D) Contact opening velocity : 10 m/s

Figure A.19: Arc current when contact opening velocity 1m/s, 4 m/s, 5.5 m/s, 10 m/s

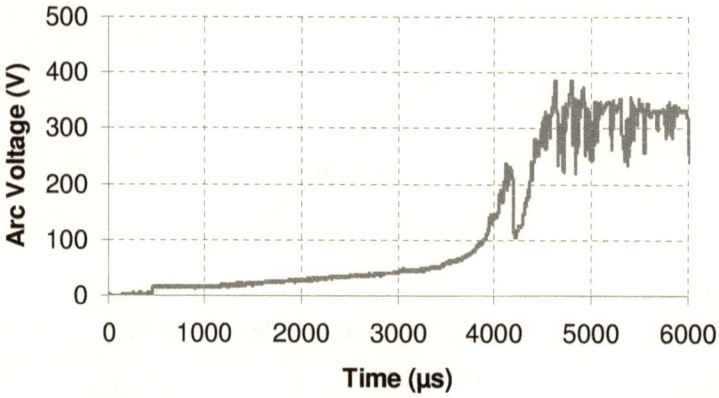

(A) Contact opening velocity : 1 m/s

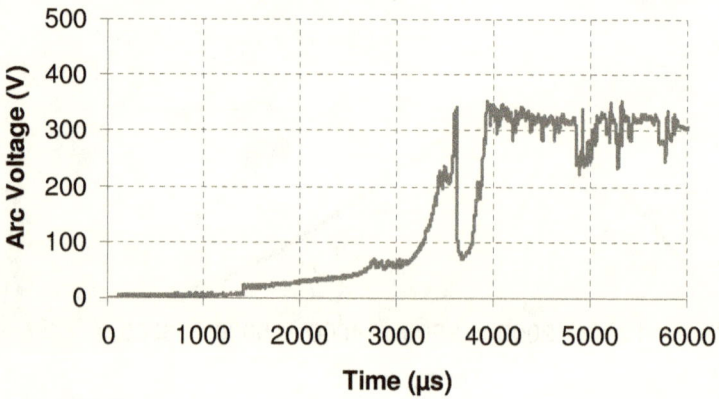

(B) Contact opening velocity : 4 m/s

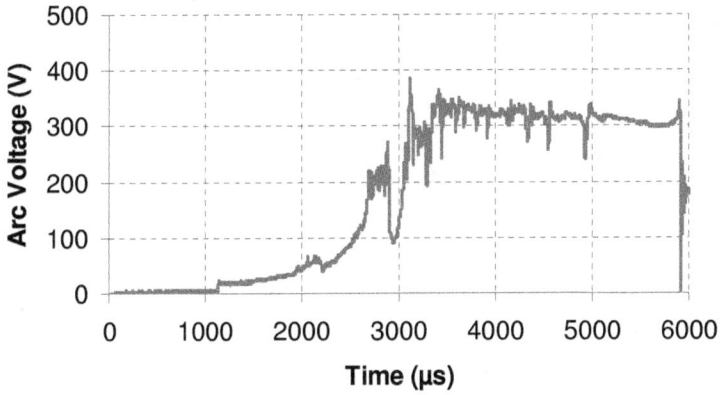

(C) Contact opening velocity : 5.5 m/s

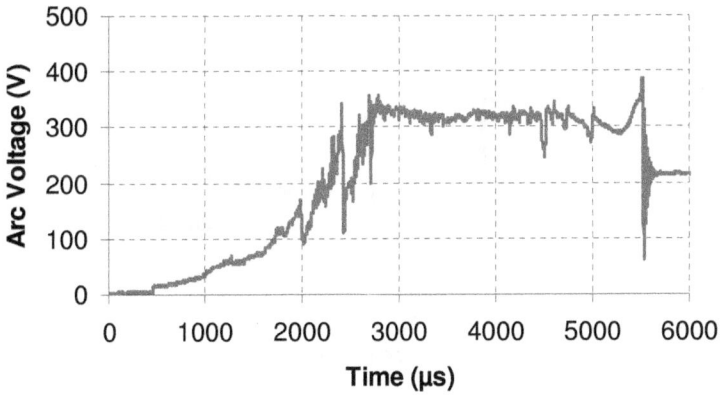

(D) Contact opening velocity : 10 m/s

Figure A.20: Arc voltage when contact opening velocity 1m/s, 4 m/s, 5.5 m/s, 10 m/s

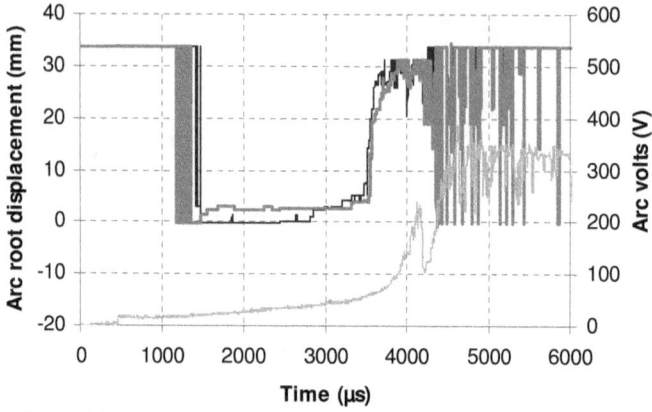

(A) Contact opening velocity : 1 m/s

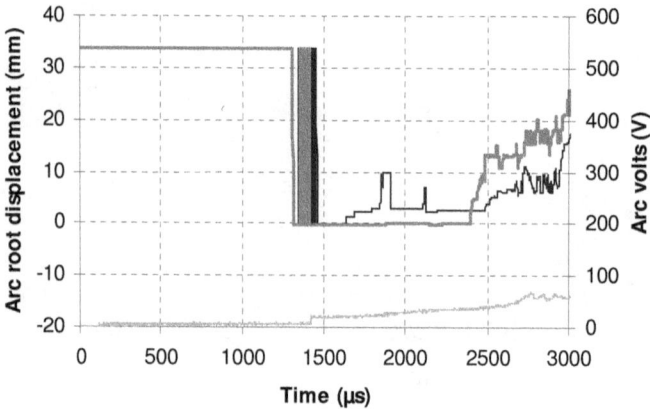

(B) Contact opening velocity : 4 m/s

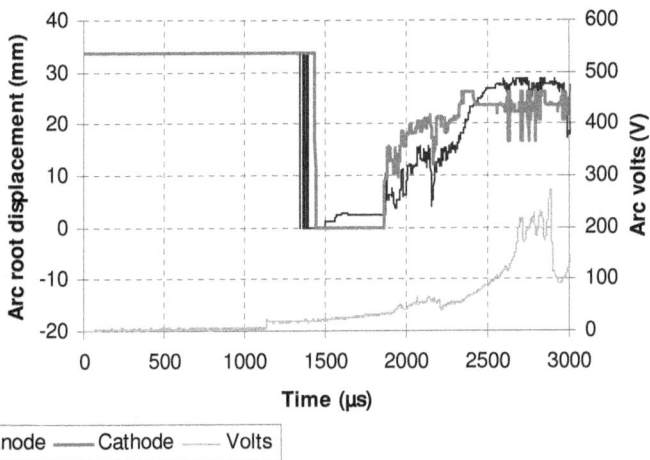

(C) Contact opening velocity : 5.5 m/s

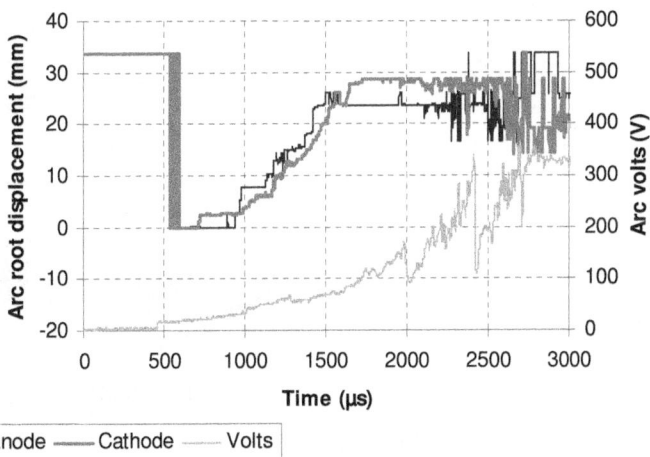

(D) contact opening velocity : 10 m/s

Figure A.21: Arc displacement when contact opening velocity 1, 4, 5.5 and 10 m/s

A.8 The Gap behind the moving contact : Opened

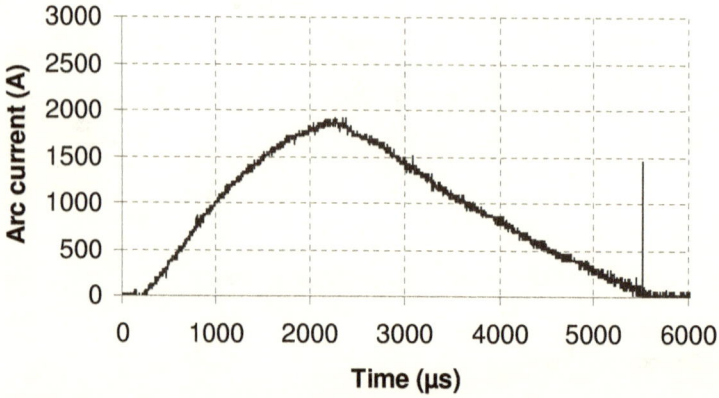

(A) Arc chamber venting : Opened

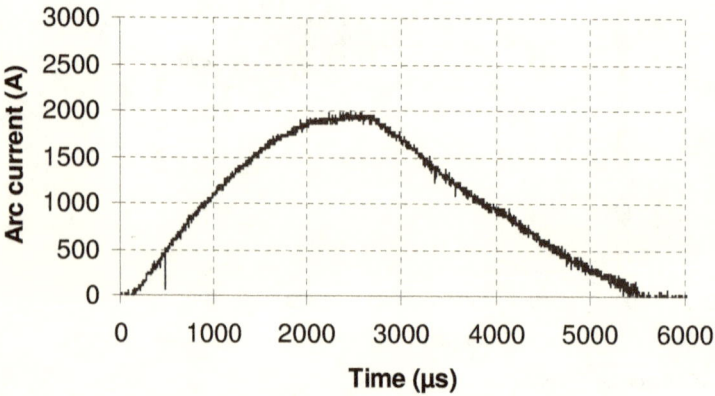

(B) Arc chamber venting : Choked

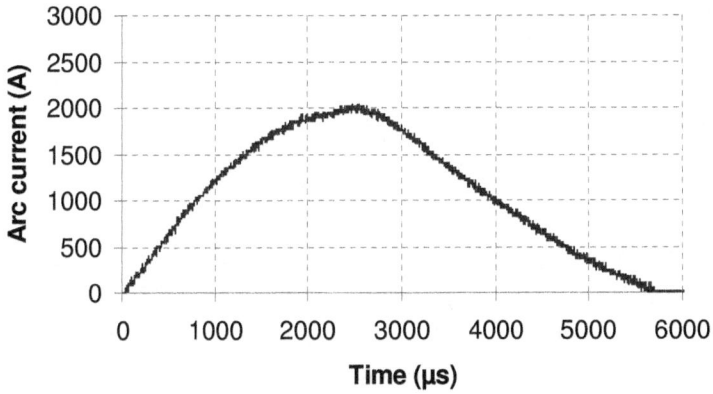

(C) Arc chamber venting : Closed

Figure A.22: Arc current when the gap behind the moving contact opened with opened, choked and closed arc chamber venting

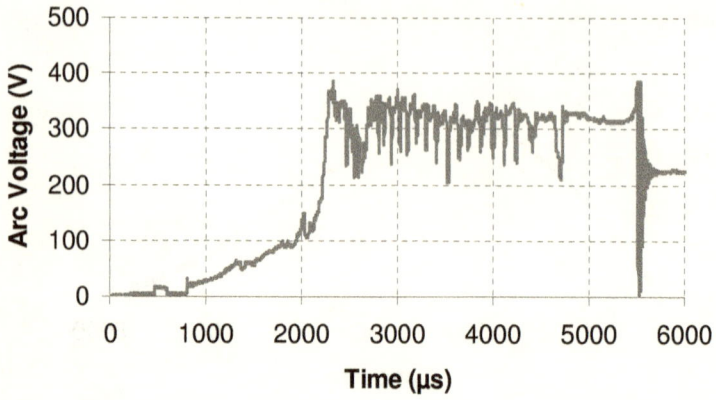

(A) Arc chamber venting : Opened

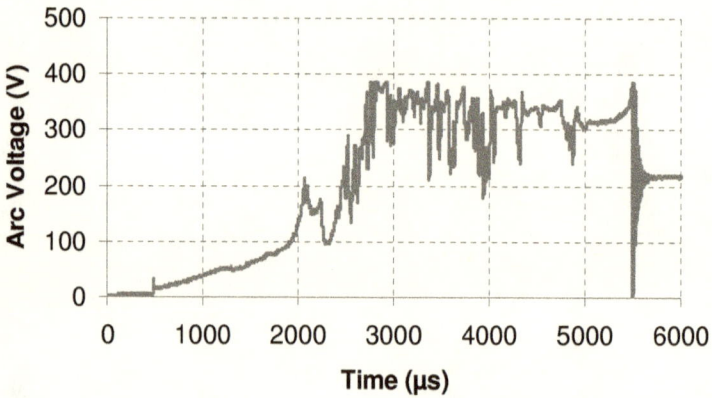

(B) Arc chamber venting : Choked

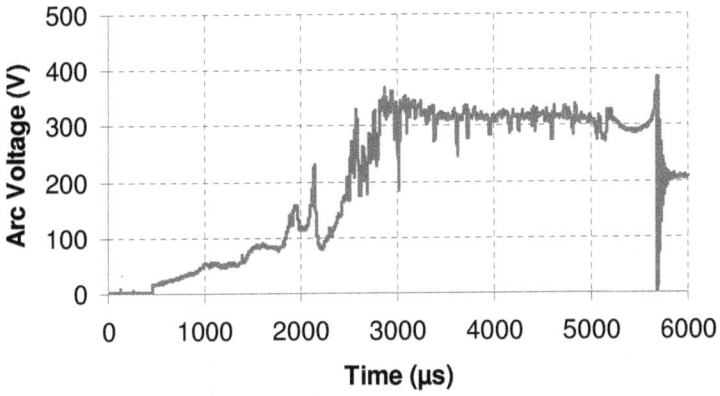

(C) Arc chamber venting : Closed

Figure A.23: Arc voltage when the gap behind the moving contact opened with opened, choked and closed arc chamber venting

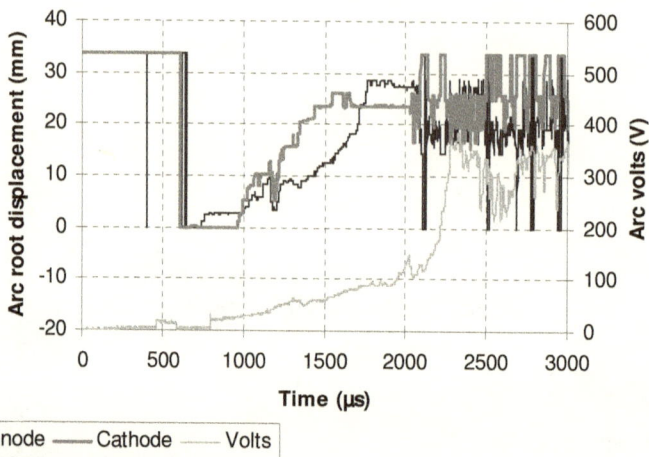

(A) Arc chamber venting : Opened

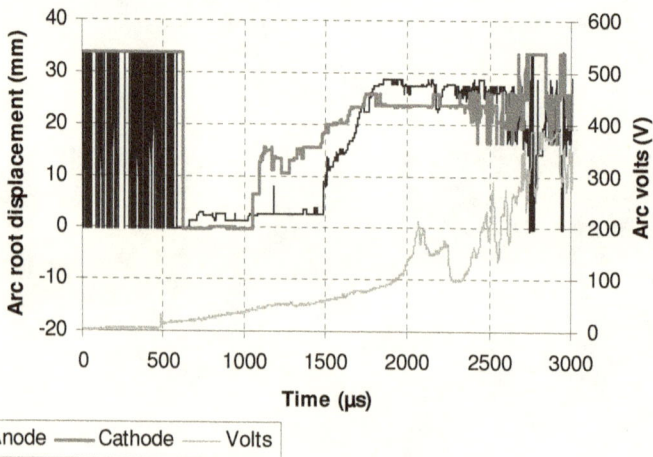

(B) Arc chamber venting : Choked

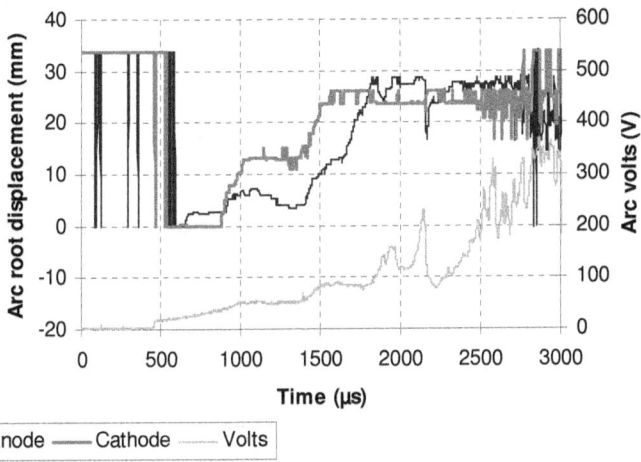

(C) Arc chamber venting : Closed

Figure A.24: Arc root displacement when the gap behind the moving contact opened with opened, choked and closed arc chamber venting

A.9 The Gap behind the moving contact : Closed

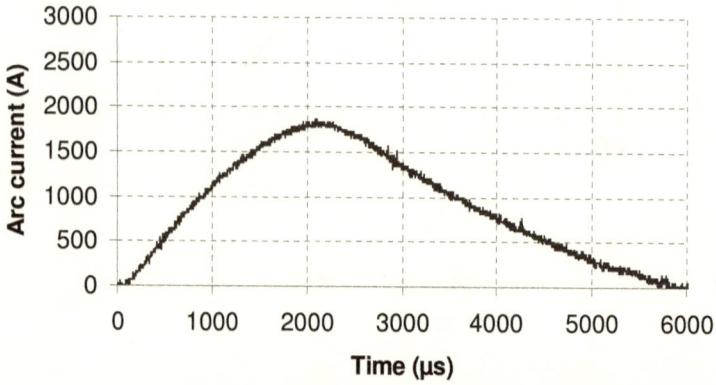

(A) Arc chamber venting : Opened

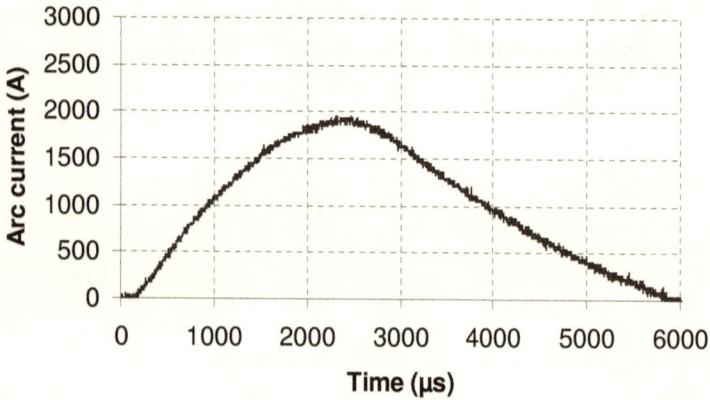

(B) Arc chamber venting : Choked

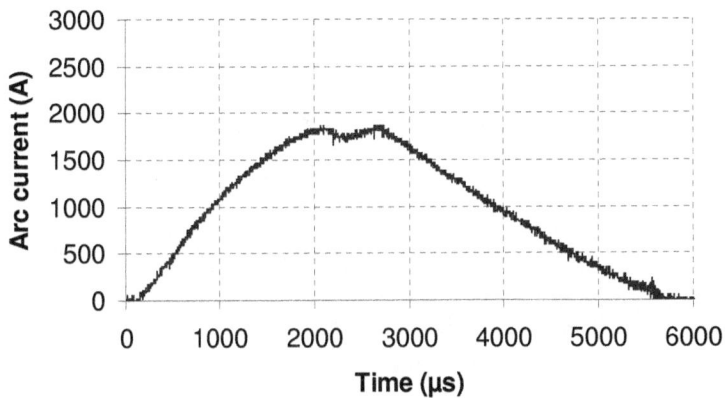

(C) Arc chamber venting : Closed

Figure A.25: Arc current with the gap behind the moving contact closed with opened, choked and closed arc chamber venting

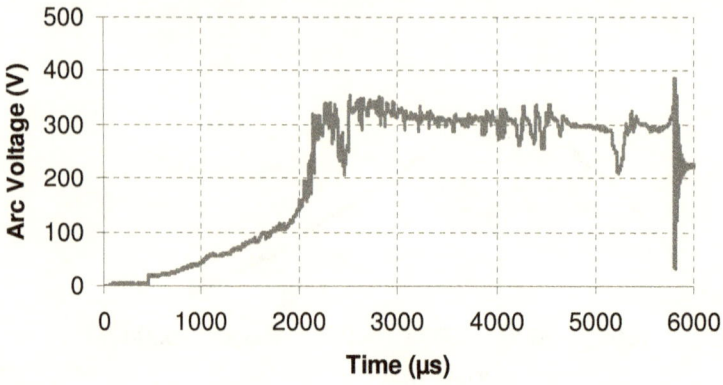

(A) Arc chamber venting : Opened

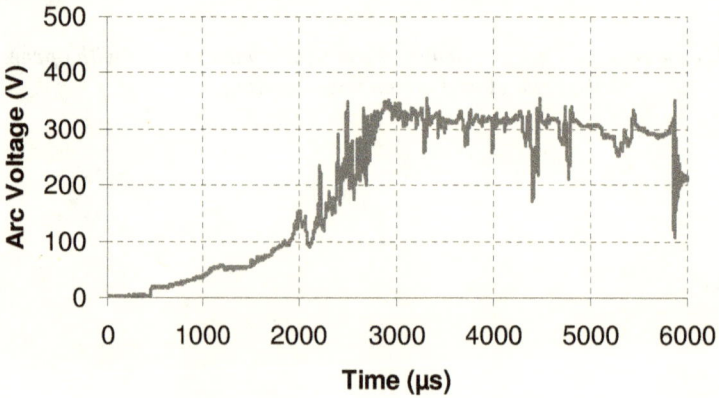

(B) Arc chamber venting : Choked

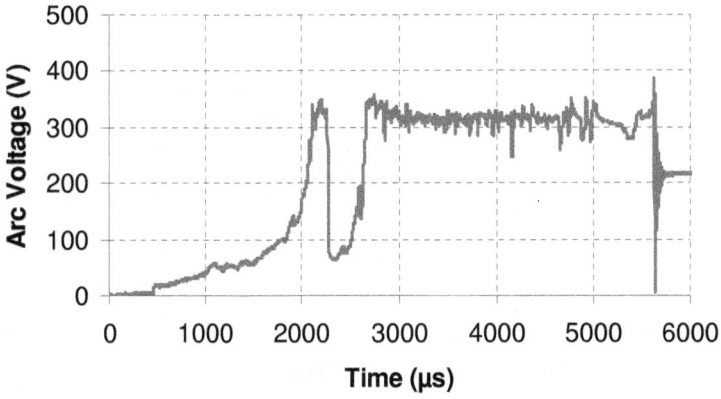

(C) Arc chamber venting : Closed

Figure A.26: Arc voltage with the gap behind the moving contact closed with opened, choked and closed arc chamber venting

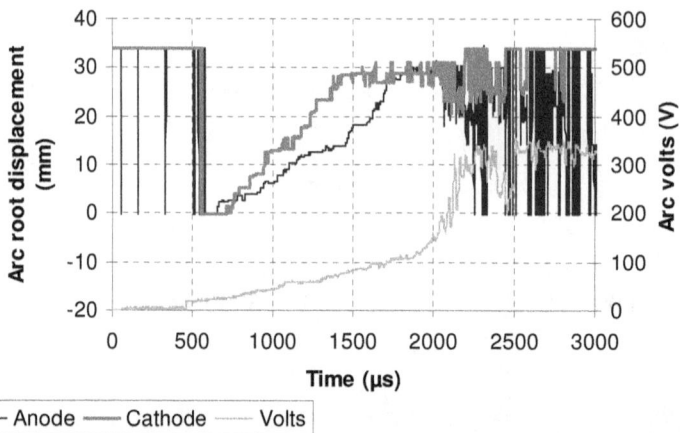

(A) Arc chamber venting : Opened

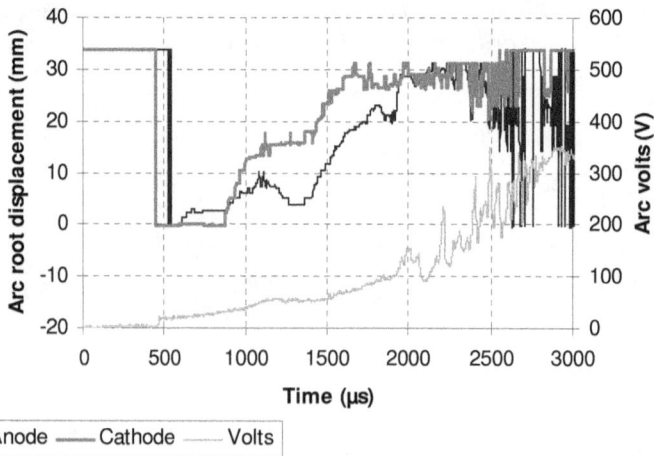

(B) Arc chamber venting : Choked

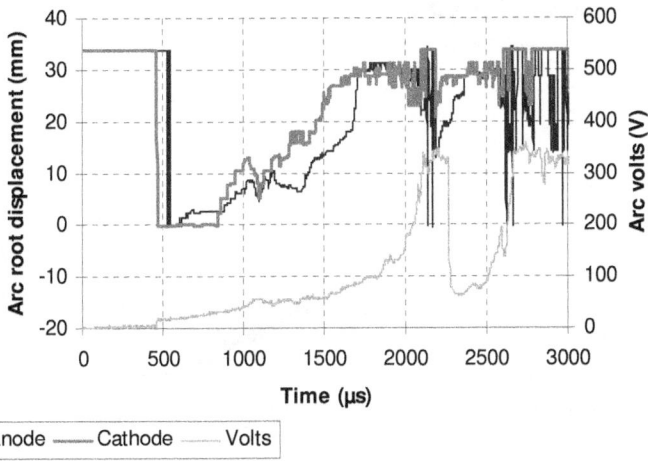

(C) Arc chamber venting : Closed

Figure A.27: Arc root displacement with the gap behind the moving contact closed and closed arc chamber venting

A.10 Arc chamber material : Polycarbonate

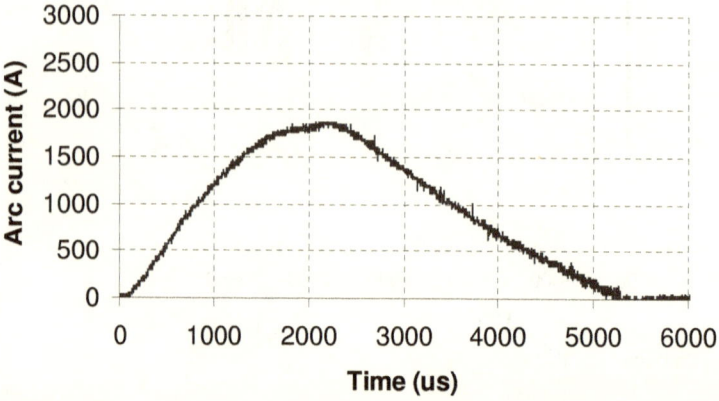

(A) Arc current

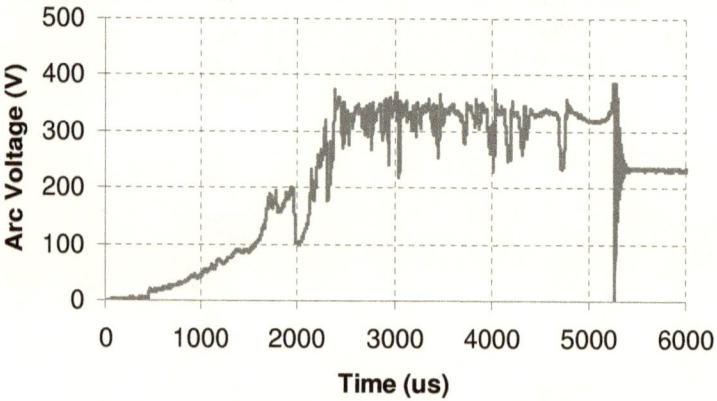

(B) Arc voltage

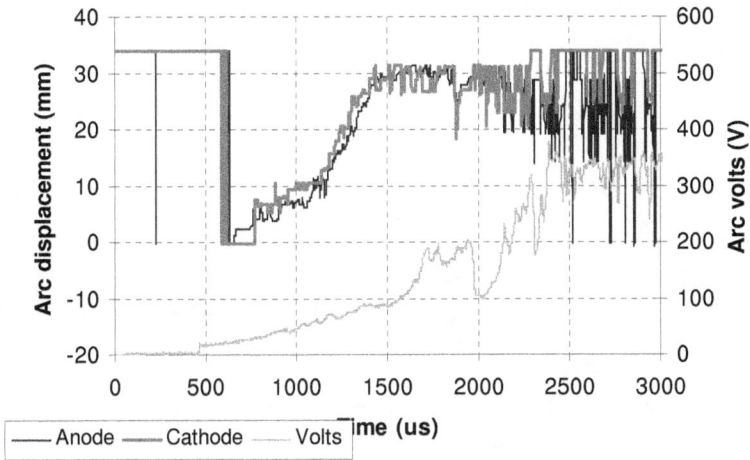

(C) Arc root displacement

Figure A.28: Arc current, arc voltage and arc root displacement of Ag/C flat contact material, 10 m/s, Anode on moving contact, 2000 A. peak current, choked, Polycarbonate arc chamber

A.11 Arc chamber material : Ceramic

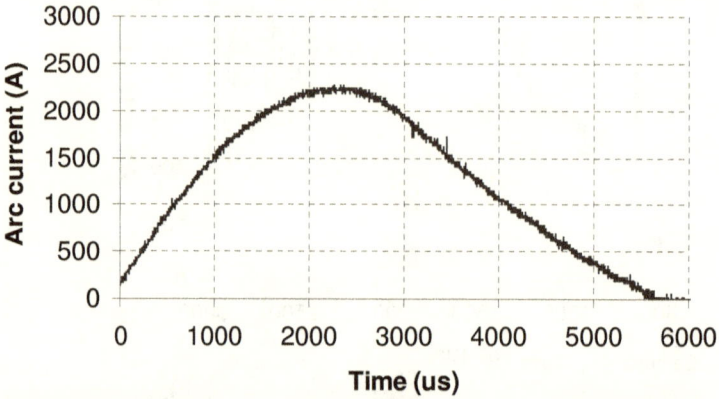

(A) Arc current

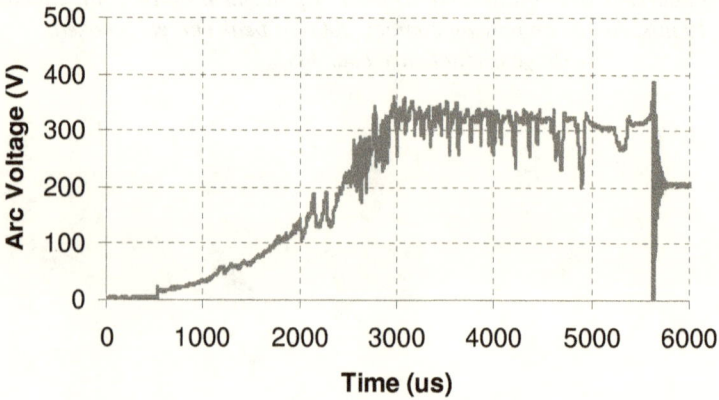

(B) Arc voltage

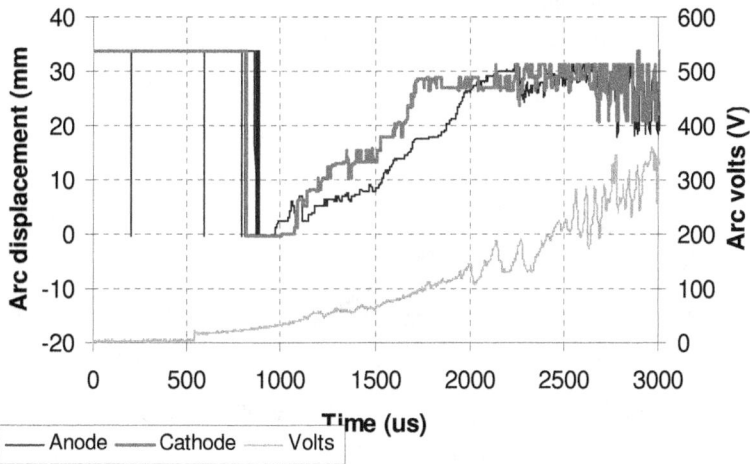

(C) Arc root displacement

Figure A.11: Arc current, arc voltage and arc root displacement of Ag/C flat contact material, 10 m/s, Anode on moving contact, choked, Ceramic arc chamber

A.12 Contact material : Ag/C step

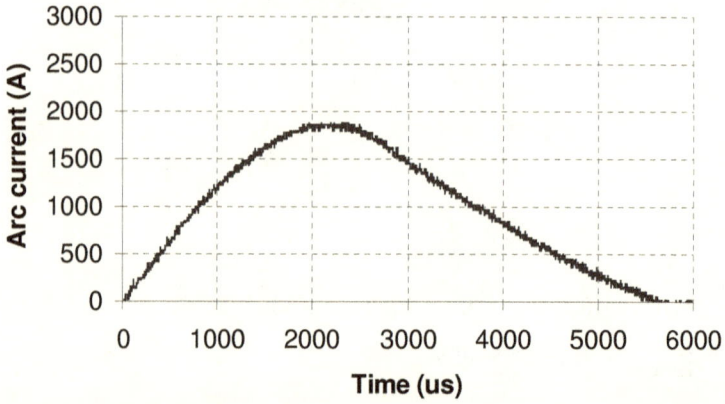

(A) Arc current

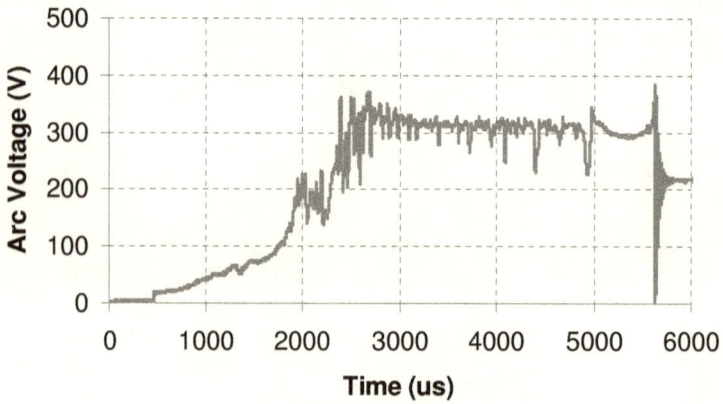

(B) Arc voltage

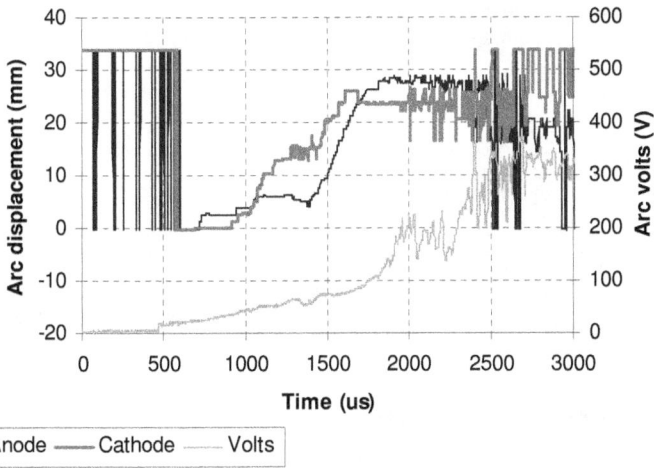

(C) Arc root displacement

Figure A.30: Arc current, arc voltage and arc root displacement of Ag/C step contact material, 10 m/s, Anode on moving contact, 2000 A. Ceramic arc chamber

A.13 Contact material : Cu punch

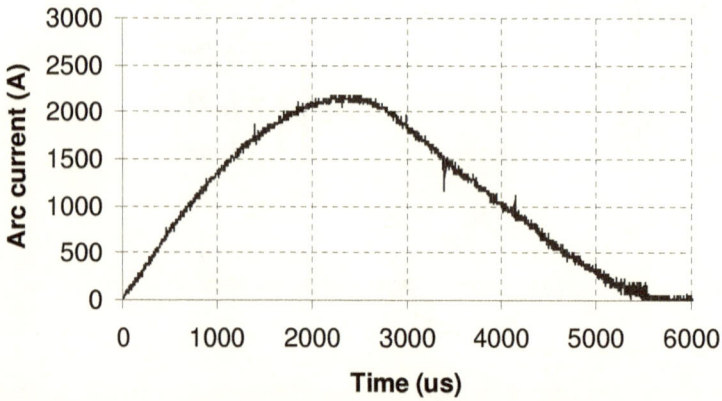

(A) Arc current

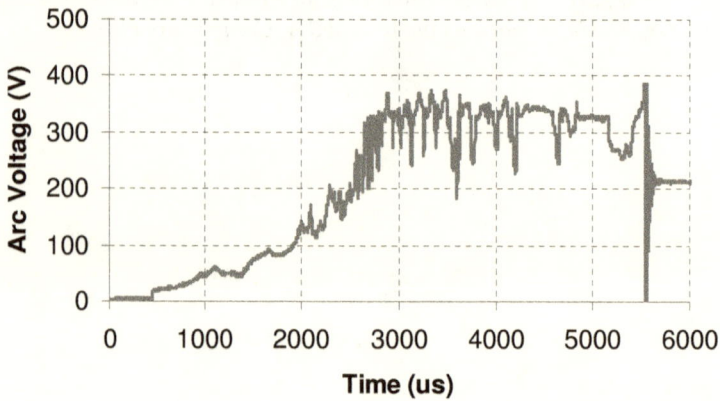

(B) Arc voltage

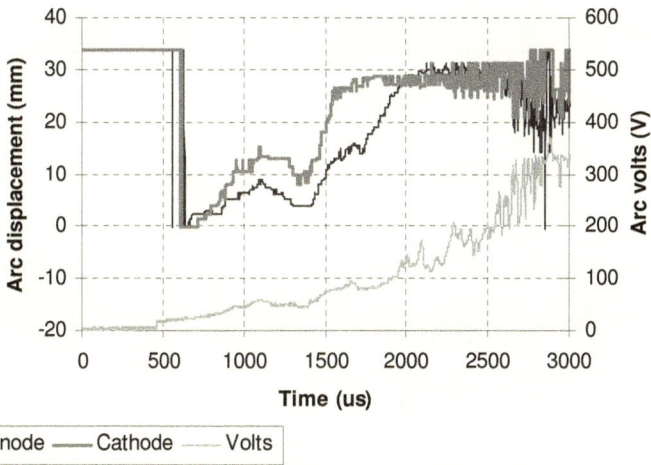

(C) Arc root displacement

Figure A.31: Arc current, arc voltage and arc root displacement of Cu punch contact material, 10 m/s, Anode on moving contact, 2000 A. peak current, choked, Ceramic arc chamber

INDEX

A

A/D converter, 84, 85, 90, 112
ablated surface, 60, 166, 177
ablating materials, 43
AC erosion, 281, 323
AC test system, 263
across the gap, 272, 288
advanced active material, 43
Ag/C step, v, 6, 7, 18, 19, 20, 21, 22, 23, 24, 25, 27, 30, 31, 103, 104, 125, 138, 154, 184, 185, 186, 188, 198, 200, 202, 205, 207, 209, 211, 212, 213, 221, 222, 244, 245, 246, 247, 248, 249, 257, 258, 269, 283, 297, 298, 299, 300, 384, 385
Ag/C step contact material, 185, 188, 246, 248, 249, 257, 258, 259, 299, 300
AIS, 3, 10
aluminium plate, 77, 89
amount of erosion, 40
amplifier circuit, 84, 89, 113
Analogue to Digital, 55
ANODE, 12, 13, 14, 15, 19, 20, 21, 22, 23, 26, 42, 45, 49, 52, 54, 57, 59, 63, 65, 70, 75, 108, 109, 116, 117, 123, 129, 130, 131, 132, 136, 137, 138, 139, 140, 141, 144, 145, 146, 147, 151, 152, 153, 155, 158, 159, 164, 165, 166, 169, 170, 171, 173, 174, 192, 193, 195, 199, 205, 206, 209, 211, 212, 213, 220, 221, 222, 260, 261, 264, 269, 273, 274, 275, 276, 281, 282, 285, 286, 287, 288, 292, 301, 315, 320, 339, 341, 343, 345, 347, 349
ANODE, 8, 16, 17, 18
anode fall, 53
ANODE ROOT, 19, 20, 21, 22, 23, 26
ANODE ROOT, 16, 17
anode root contact time, 108, 130, 131, 132, 137, 155, 159
anode root on the fixed contact, 139
anode surface, 53
ARC CHAMBER, 4, 5, 7, 9, 11, 12, 13, 14, 15, 16, 17, 18, 19, 20, 21, 22, 23, 25, 26, 29, 39, 41, 43, 44, 45, 46, 48, 49, 54, 55, 56, 57, 58, 59, 61, 62, 63, 65, 66, 67, 68, 69, 70, 71, 75, 77, 80, 82, 83, 87, 89, 90, 92, 100, 105, 107, 108, 109, 110, 111, 112, 113, 118, 119, 120, 121, 129, 130, 131, 132, 136, 137, 138, 140, 141, 144, 145, 146, 147, 151, 153, 155, 158, 159, 160, 161, 162, 163, 165, 166, 167, 168, 169, 170, 171, 172, 173, 174, 175, 177, 178, 179, 181, 182, 185, 186, 187, 188, 189, 190, 192, 204, 205, 206, 207, 209, 211, 212, 213, 214, 217, 220, 221, 223, 260, 261, 263, 265, 267, 269, 272, 274, 276, 278, 279, 283, 284, 285, 287, 288, 289, 290, 291, 303, 304, 305, 306, 307, 308, 309, 310, 315, 317, 326, 327, 328, 329, 331, 339, 341, 343, 345, 347, 349, 369, 371, 373, 375, 377, 379, 381, 383, 385, 387
Arc chamber, 2, 4, 5, 11
arc chamber geometry, 68, 83, 153
ARC CHAMBER MATERIAL, 11, 12, 26, 29, 44, 48, 49, 67, 68, 69, 78, 100, 129, 130, 131, 136, 146, 161, 166, 177, 283
Arc chamber material, 2, 4, 5
arc chamber outlet, 161
Arc chamber sidewalls, 77
arc chamber surfaces, 68
arc chamber venting, 62, 68, 71, 131, 132, 136, 137, 138, 140, 144, 147, 151, 153, 155, 159, 160, 162, 211, 212, 261, 269, 309
Arc chamber venting, 61, 101, 126, 127, 128, 129, 133, 134, 135, 136, 154, 156, 157, 158, 162, 167, 184, 324, 326, 327, 328, 329, 330, 331, 332, 333, 334, 335, 336, 337, 338, 339, 340, 341, 342, 343, 344, 345, 346, 347, 348, 349, 356, 368, 369, 370, 371, 372, 373, 374, 375, 376, 377, 378, 379
arc column, 45, 52, 57, 289, 306
arc commutation delay, 63, 269
arc concept, 47
arc control, 43, 72, 82
ARC CONTROL, 23, 24, 25
arc crosses the opened gap, 288
arc current, 45, 53, 63, 66, 67, 68, 72, 105, 108, 113, 137, 145, 193, 197, 198, 202, 210, 211, 213, 222, 262, 263, 264, 265, 268, 269, 271, 272, 273, 277, 281, 282, 287, 301, 308, 309, 311, 324, 351, 353, 355
arc current at the point, 263
arc current lead, 63
arc dimension, 204
arc dynamics, 56, 65
arc energetics, 260
arc energy, 209
arc flow velocity, 205
arc gases, 283
arc heat flux, 63
arc heating and melting, 276
arc ignited, 287
arc ignition, 56, 69, 113
arc image contour movies, 108
arc images movies, 107
arc immobility, 44, 46, 56, 58, 59, 60, 166, 214, 278
arc length, 198, 202, 204, 270, 271
arc literature, 51
arc motion, 41, 42, 43, 44, 45, 47, 51, 55, 56, 60, 62, 66, 67, 68, 70, 71, 78, 108, 110, 119, 122, 172, 193, 260, 265, 266, 267, 269, 272, 277, 305, 314, 318
arc movement, 56, 59, 62, 86, 317, 322
arc movie, 42
arc power, 46, 68, 108, 210, 211, 213, 218, 219, 220, 221, 222, 276, 277, 278, 301
ARC ROOT, 4, 11, 12, 19, 20, 21, 25, 26, 42, 43, 44, 45, 46, 47, 48, 49, 51, 56, 57, 58, 59, 62, 65, 66, 70, 71, 72, 73, 75, 76, 99, 100, 105, 107, 108, 109, 110, 111, 113, 115, 116, 117, 118, 120, 121, 123, 124, 125, 126, 129, 131, 132, 136, 137, 138, 139, 141, 144, 145, 146, 147, 151, 152, 153, 159, 160, 161, 162, 163, 166, 168, 172, 173, 174, 175, 176, 177, 181, 184, 188, 189, 190, 192, 193, 194, 198, 199, 200, 201, 202, 203, 206, 209, 210, 211, 212, 213, 214, 220, 221, 222, 260, 261, 262, 263, 264, 265, 266, 267, 268, 269, 270, 271, 272, 273, 274, 275, 276, 277, 278, 279, 280, 281, 285, 286, 287, 288, 289, 290, 300, 301, 302, 303, 304, 305, 306, 307, 308, 309, 310, 311, 314, 324, 381, 383, 385, 387
Arc root, 4, 6, 10, 11, 12, 13, 14, 15
arc root commutation, 212, 305
arc root contact time, 44, 47, 58, 62, 65, 72, 109, 115, 116, 117, 118, 121, 123, 126, 131, 132, 137, 138, 139, 145, 146, 147, 153, 159, 160, 161, 162, 172, 175, 189, 213, 261, 264, 265, 266, 267, 269, 277, 285, 303, 304, 305, 306, 307, 308, 309, 310, 311, 324
arc root displacements, 159
arc root mobility, 43, 44, 59, 62, 76, 184, 188, 260
arc root movement, 308
arc root moving from contact region, 44
arc root to move from the contact region, 311
arc roots, 40, 42, 57, 75, 164, 166, 175, 176, 260, 284
arc spectrum, 44, 49, 72, 93, 112, 119, 120, 161, 177, 179, 181, 182, 183, 184, 190, 303

gas dynamic model, 274
gas dynamic modelling, 72, 205, 210
gas dynamics, 311, 314, 323
gas flow, 41, 44, 46, 47, 48, 58, 60, 66, 67, 69, 71, 72, 99,
    166, 172, 192, 203, 209, 215, 260, 273, 274, 275, 288,
    289, 303
gas pressure, 87, 119, 172, 204, 260, 303
gas temperature, 52
gases affects, 43
gassing materials, 61, 68
geometry of contact, 131, 160
geometry of the contact and conductors, 311
governing factor, 46, 214
graphite, 39, 59, 65, 319

H

heat liquid, 218
heat metal, 218
high contact opening velocities, 301
high electric field, 51
HIGH SPEED ARC IMAGING SYSTEM, 3
high speed Arc Imaging System (AIS), 42, 55, 71, 75, 260
high voltage, 39, 54, 70
high-speed film, 55
hot gases, 111, 278, 288
hydrogen content, 61
hydrogen molecule, 60

I

Ideal gases constant, 215
independent of the material, 262
influence of the arc, 48, 110, 145, 160, 189, 285, 309, 310
initial time, 43
integration time, 91, 112, 310
interchangeable components, 110
ion bombardment, 52
IONISATION, 22
ionised gas, 289
ionized material electrostatic, 289

J

Joule heating, 46

L

light intensity, 55, 114, 115, 121, 274
light intensity distribution, 121, 275
Lorentz force, 46, 214
low burning, 41
Low contact velocity, 43
low light levels, 121, 275
low mass of anodic gases, 290
low voltage switching devices, 43
lower contact opening velocities, 43
lower velocities, 284, 301
lowest temperature rise, 249
luminescence, 64

M

Macor machineable ceramic, 77
magnetic, 36, 37, 39, 45, 47, 49, 51, 54, 56, 58, 59, 63, 65, 66,
    67, 68, 69, 70, 72, 97, 137, 184, 192, 193, 194, 195, 196,
    197, 198, 199, 201, 202, 210, 222, 262, 264, 265, 269,
    270, 271, 272, 273, 274, 275, 276, 279, 301, 305, 307,
    308, 311, 319, 322, 323

magnetic driving force, 66, 271
magnetic field, 59, 65, 66, 193, 271, 307
magnetic flux density, 66, 193, 271, 275
Magnetic flux density, 215
magnetic force, 137, 193, 307
magnetic forces, 36, 37, 45, 47, 66, 69, 72, 192, 193, 194,
    197, 198, 199, 201, 202, 210, 222, 270, 271, 272, 273,
    274, 276, 279, 301, 305, 307, 308, 311
Magnetic potential, 196
Magnitude of distance, 196
main structure, 57
Mass, v, 5, 6, 23, 24, 27, 30, 31, 218, 227, 228, 229, 232, 233,
    236, 237, 238, 241, 242, 243, 246, 247, 248, 251, 252,
    253, 255, 256, 258, 278, 293, 294, 296, 298, 299
mass flow, 46, 47, 72, 192, 219, 220, 221, 223, 278, 279, 280,
    281, 282, 283, 290, 301, 304, 307
mass flow rate, 46, 72, 219, 220, 221, 223, 279, 280, 307
mass per mole, 215, 217
material transfer, 281, 282
maximum intensity, 179, 183, 186, 188
maximum Mass, 229, 234, 243, 253, 257, 258, 296, 299
maximum pressure, 169, 170, 171, 175, 176, 190
maximum short circuit current, 145
maximum temperature rise, 229, 234, 244, 254, 256, 257,
    258, 294, 297, 300
MCB, 2
measure the spectral data, 90, 119, 303, 310
mechanisms, 43, 217
melting point, 41, 215, 216
metal atoms, 54
metal vapour, 59, 61, 71, 215
metallic vapour, 64, 71, 176, 181, 283
methodologies and techniques, 47
miniature circuit breaker, 38, 39, 40, 49, 55, 77, 82, 118, 274,
    303, 314, 317
Miniature Circuit Breaker, 42
minimum contact gap, 262, 305
minimum energy, 227, 232, 241, 255, 256, 293, 296
minimum forces, 304, 311
minimum gap, 63, 262
minimum Mass, 238, 243, 253, 255, 256, 258, 294, 296, 299
minimum temperature rise, 229, 239, 254, 256, 257, 297, 300
modeling of magnetic forces, 45
modelling results, 47, 198, 202
molecules and ions, 181
molten bridge, 53, 64
motion of the arc root, 290
movement of the arc root, 42, 172, 307
moving arc runner, 70, 92, 263
MOVING CONTACT, 3, 4, 6, 7, 10, 12, 13, 14, 15, 16, 17, 18, 20,
    21, 22, 23, 26, 29, 39, 44, 45, 48, 56, 57, 70, 77, 92, 99,
    100, 101, 104, 108, 109, 111, 113, 115, 116, 117, 119,
    123, 125, 126, 127, 129, 130, 131, 132, 134, 136, 137,
    138, 139, 140, 144, 145, 146, 151, 152, 153, 154, 155,
    158, 159, 160, 161, 162, 163, 164, 165, 167, 169, 170,
    171, 172, 173, 174, 175, 177, 178, 180, 184, 185, 187,
    189, 192, 193, 195, 198, 199, 205, 206, 209, 221, 222,
    267, 273, 276, 281, 283, 285, 286, 287, 288, 289, 290,
    291, 301, 304, 306, 307, 310, 324, 326, 327, 329, 331,
    344, 345, 347, 349, 350, 368, 374, 381, 383, 385, 387
moving contact arc runner, 287, 289, 306

N

nitrogen, 36, 61, 65, 189, 282, 284
numerical data, 115

O

one-dimensional, 274

open gap, 163, 287, 290, 301
opening contacts, 39, 54, 58, 62, 69, 137, 287
operating pressure, 87
OPERATION OF MCB, 2
optical data, 42, 44, 46, 55, 70, 75, 82, 115, 118, 123, 192, 206, 210, 260, 266, 267, 274, 303, 308
optical fibre, 43, 55, 62, 77, 82, 84, 90, 91, 92, 93, 112, 119, 124, 303, 310, 314
optical fibre imaging and analysis system, 62
optical fibre spectrometer, 43, 90, 91, 93, 303, 310
optical system, 260, 264
optical techniques, 55
Organic gas components, 61
organic vapours, 60
oxygen, 36, 61, 65, 176, 189, 282, 284

P

parallel runners, 287
parameters, 42, 43, 47, 75, 93, 100, 105, 116, 222, 260
Peak Limited current, 38
PEAK LIMITED CURRENT, 8
Permeability, 196
photodiode, 84
photo-transistors, 86
physical mechanism, 277
physical principles, 45, 222
piezo resistive, 44, 87, 119, 303
plain Cu, 125
Planck's constant, 224
plasma, 39, 43, 52, 53, 60, 62, 67, 69, 71, 166, 177, 283, 289, 314, 315, 319, 320
plasma chemical reactions, 69
plasma compositions, 67
point at which the arc moves, 269
POINT ON WAVE, 8
Point on Wave (POW)., 38
polycarbonate, 101, 110, 166, 168, 171, 173, 175, 177, 178, 179, 181, 182, 189, 190, 284
Polycarbonate, v, 5, 6, 16, 17, 23, 27, 30, 104, 167, 168, 169, 173, 174, 225, 226, 227, 228, 229, 230, 255, 267, 292, 293, 294, 325, 380, 381
polycarbonate arc chamber, 172, 173, 175, 178, 190, 284
Polymer fibre, 82
polymer vapour, 217
position of the Centre of Intensity (COI), 115
power dissipation, 289
power thyristor, 114
powerful tool, 279
pressure, 44, 45, 46, 47, 48, 56, 58, 60, 61, 62, 63, 65, 67, 68, 69, 71, 75, 87
pressure across the arc, 205, 274, 275
pressure behind the moving contact, 48, 163, 164, 165, 166
pressure build up, 166, 274
pressure data, 275, 303
Pressure effects, 44
pressure in the arc chamber, 44, 68, 161, 162, 166, 172, 177, 303
pressure in the gap, 44, 45, 164, 165
pressure measurements, 260
pressure ratio, 274, 275
pressure sensor, 87, 113, 119
pressure transducer circuits, 87
pressure transducers, 87, 113, 119, 162, 260, 303, 306
properties of plasmas, 60
Prospective Peak Short circuit current, 37

Q

Qbasic computer programme, 107

quantity of the flow, 287, 309, 310
quartz glass, 109, 110

R

radius of the arc, 68, 161
rapid opening, 43
rate of flow, 278
ratio of pressure, 205
real arc root lengths, 193
reflectance and emission, 93
Relative permeability, 196
resistance of the arc, 54
RLC circuit, 63

S

same chemical element, 256, 257, 258, 294, 296, 300
second high Mass, 238, 256
semi-empirical model, 45
shape of the arc, 59, 121, 274
shock front, 203
shock wave, 56, 203, 274, 275
short arc, 46, 193, 209, 214
short circuit, 35, 36, 37, 38, 39, 41, 44, 48, 49, 54, 55, 56, 57, 58, 63, 65, 67, 68, 69, 70, 72, 77, 78, 100, 107, 110, 111, 113, 118, 120, 121, 125, 129, 130, 137, 138, 139, 140, 141, 144, 145, 146, 151, 155, 158, 160, 161, 177, 189, 198, 261, 263, 265, 269, 273, 291, 303, 304, 307, 308, 313, 314, 357, 359, 361
SHORT CIRCUIT CURRENT, 8, 11, 12, 13, 14, 15, 26, 37, 38, 39, 41, 70, 110, 139, 141, 144, 145, 307
SHORT CIRCUIT CURRENT, 2, 4, 7
short circuit fault, 38, 39, 54, 69, 137
simulate, 42, 57, 70, 77, 118, 194, 260, 303
smart material actuation, 302
solenoid trip mechanism, 39
sound speed, 204
specific heat, 215, 216
Specific heat ratio, 215
Spectra, 90, 91, 178, 179, 180, 182, 183, 185, 186, 187, 188
spectra line, 183, 188
spectral measurements, 46
spectrometer, 3, 9, 47, 49, 75, 90, 91, 92, 93, 94, 111, 112, 119, 182, 184, 260, 261, 284
SPECTROMETER, 3, 4
spectrometer channels, 93
spectrometer measures, 90
spectroscopic, 64, 93
spherical contact, 58
standard deviation, 109, 111, 123
steel backing plate, 108
stepping motor, 119
steps and gaps, 286
streak camera, 55
STRUCTURE OF MINIATURE CIRCUIT BREAKER, 8
supply polarity, 48, 49, 100, 120, 131
surface topology, 59
switching arc, 45, 314

T

temperature, 36, 41, 52, 53, 60, 62, 68, 69, 87, 166, 177, 203, 204, 215, 216, 217, 276, 320, 321
Temperature, 215
temperature rise, v, 224, 229, 234, 239, 244, 249, 254, 256, 257, 258, 294, 297, 300, 301, 309
TEMPERATURE RISE, 5, 224

theories relating gas, 215
Thermal, 3, 5
thermal conditions, 36
thermal effects, 215
thermal forces, 54, 70
thermal properties, 82
thermally driven flow, 46, 214
thermally enhanced field emission (T-F emission), 51
thermodynamic calculation, 279
thermodynamic model, 279
thermodynamic properties, 60
thickness of the conductor, 202
thin silicon diaphragm, 87
total mass flow, 72, 279
trigger, 78, 106, 112, 113, 114, 310
typical waveforms, 211

U

upper trace, 116

V

values of arc current, 63, 269
vaporized, 46, 63, 214, 283
vapour layer, 217

vapourised material, 289
variation, 212, 213
Vector fields, 196, 313
velocity of the anode root, 274
velocity of the arc root., 308
velocity of the cathode root, 274
velocity of the flows, 206
vent in the arc chamber, 153, 162
venting area, 61, 71, 132, 153, 287
vibration, 89
voltage drop, 41, 52, 97
volume flow of the arc gas, 304
volume flow rate, 46, 72, 219, 220, 221, 223, 307
volume of gas, 278

W

wall vaporization, 287
wavelength, 30, 31, 64, 90, 93, 94, 111, 179, 181, 183, 186, 188, 189, 224, 225, 226, 227, 229, 230, 231, 232, 234, 235, 236, 237, 238, 239, 240, 241, 244, 245, 246, 247, 249, 250, 251, 253, 254, 255, 256, 257, 258, 259, 283, 293, 299, 300
wavelength of spectral line, 94
weightless piston, 203